Intermediate
Social
Statistics

Intermediate
Social
Statistics

A Conceptual and Graphic Approach

Robert Arnold

OXFORD
UNIVERSITY PRESS

OXFORD
UNIVERSITY PRESS

Oxford University Press is a department of the University of Oxford. It furthers the University's objective of excellence in research, scholarship, and education by publishing worldwide. Oxford is a registered trade mark of Oxford University Press in the UK and in certain other countries.

Published in Canada by
Oxford University Press
8 Sampson Mews, Suite 204,
Don Mills, Ontario M3C 0H5 Canada

www.oupcanada.com

Library and Archives Canada Cataloguing in Publication

Arnold, Robert, 1943-, author
Intermediate social statistics : a conceptual and graphic
approach / Robert Arnold.

Includes bibliographical references and index.
ISBN 978-0-19-901207-7 (pbk.)

1. Social sciences--Statistical methods. 2. Statistics.
I. Title. II. Title: Social statistics.

HA29.A76 2014   300.72'7   C2014-904760-6

Cover image: ©iStockPhoto.com / handhead

Oxford University Press is committed to our environment.
This book is printed on Forest Stewardship Council® certified paper and comes from responsible sources.

Printed and bound in Canada

1 2 3 4 — 18 17 16 15

Key Formulae

1. **The mean**
   $\bar{x} = \Sigma x_i / N$

2. **The "entropy" measure of dispersion**
   $- \Sigma p_i[\log_2(p_i)]$

3. **The Inter-quartile Range**
   $Q3 - Q1$ (or $P_{75} - P_{25}$)

4. **The Inter-decile Range**
   $D9 - D1$ (or $P_{90} - P_{10}$)

5. **The standard deviation,**
   $[\Sigma (x_i - \bar{x})^2 / N]^{.5}$ or $\sqrt{\Sigma (x - \bar{x}_i)^2 / N}$

6. **A standardized variable**
   $(x_i - \bar{x}) / SD(x)$

7. **Standard Error of $\bar{x}$**
   $SD(x) / N^{.5}$, or $SD(x) / \sqrt{N}$

8. **Standard Error of p**
   $[p(1 - p) / N]^{.5}$

9. **Chi-square**
   $X^2 = \Sigma (O - E)^2 / E$

10. **Phi (φ)**
    $\sqrt{X^2 / N}$

11. **The General PRE Formula**
    $(\text{Error 1} - \text{Error 2}) / \text{Error 1}$

12. **Lambda (λ)**
    $(\Sigma m_j - M) / (N - M)$

13. **Gamma (γ)**
    $(C - D) / (C + D)$

14. **Q**
    $(ad - bc) / (ad + bc)$

15. **Somers' d**
    $(C - D) / (C + D + T_y)$

16. **A Standardized Residual**
    $\pm [(O - E)^2 / E]^{.5}$

17. **The Odds Ratio (OR)**
    ad / bc

18. **The Covariance**
    $\Sigma (x_i - \bar{x})(y_i - \bar{y}) / N$

19. **Pearson's r**
    $Cov(x,y) / [SD(x)SD(y)]$

20. **Partial Q**
    $(\Sigma ad - \Sigma bc) / (\Sigma ad + \Sigma bc)$

21. **Bivariate b**
    $Cov(x,y) / Var(x)$

Brief Contents

Detailed Contents

PART IV   EXAMINING CROSSTABULATIONS 167

PART V   REGRESSION 197

Preface

This book stems from 19 years of teaching undergraduates in sociology. In my first year of teaching I wrote material to supplement the text I was using. In my second I replaced sections that students had trouble following. Encouraged by their response, I wrote material to cover topics not treated in the text I had moved to. Six years ago I decided to teach entirely from my own material. This has been revised into the volume you see. Apart from material created in light of reviewers' comments, everything been class tested, most of it at least five times.

The book moves quickly over topics that are standard in a one-semester first course. On these themes it provides only a review or a review with added sophistication. It then moves to cover a range of material broader than that of texts often used with Canadian undergraduates. This may be readily seen in the detailed Table of Contents. Notice, in particular, the extended section on regression, the analytic strategy most common in the journals today.

Since a second course may vary to suit the research interests of a department, and log-linear models are very important in some areas of sociology, a chapter on basic log-linear modelling is available for download from the course website, along with other instructional materials. In my experience, students can handle this chapter as well as other material in the latter half of the book.

Although this book does not emphasize calculation, it does present numerous formulae. When it does so, it always explains why the formula is what it is. For example, students have often been asked to memorize the formula for a standard deviation, or of Pearson's r, or to do calculations based on these formulae, without knowing why they take the form they do. I have included a rationale for each. In the explanations I have not used mathematics beyond high school algebra, but where a bit of algebra could be helpful I have been happy to use it.

Broad coverage and detailed explanation can be combined in a text intended for a one-semester course because of a conceptual approach, together with extensive use of graphics. By a conceptual approach, I mean simply one focused on why we use particular statistics, what they can tell us, how they can mislead us, and what to do to prevent being misled (or misleading others). Within this framework, students may be asked to work out the values of basic statistics, but primarily as an aid to conceptual understanding. Since computers now do almost all serious statistical calculation, both in academia and outside, we have reduced reason to emphasize do-it-yourself calculation, and can therefore move students further in their conceptual understanding.

This is made much easier by frequent use of graphics. In this respect, I have been much influenced by graphic methods often used in the sub-discipline of statistical graphics. In one case, that of the chapter on conditional tables, I can move through the material in less than half the time I needed before adding graphs. Apart from its advantages in teaching, graphical communication is common in the leading journals of sociology, in the business world, in reporting research results to general audiences, and increasingly in the mass media. Exposing students to a wide range of graphics is therefore of much more than instructional advantage.

Each chapter concludes with a set of review questions. I have used many of these, or variations on them, as short answer questions in examinations. If I present a question containing data or statistics, the numbers are never those in the handbook, but if students understand the answers to the review questions, and have not just tried to memorize right answers, they have acquired the key knowledge I hope to pass on. If examinations take another form, working out the answers to the review questions should provide excellent study notes.

While I have made no compromises in content, apart from omitting material requiring matrix algebra, I have aimed to keep the writing simple. Of course technical terms must be used, but otherwise I have aimed at a level which university-bound students in the later years of high school could readily handle.

In the interests of clear exposition, I have sometimes explained a point with an artificial example, then moved to one based on real data. In other cases, real data has lent itself well to purposes of exposition. In these cases I have often presented a second example, based on the experience that a second example has helped students to feel secure in their understanding. I have also been influenced by a student observation that, when studying, if a point is clear from the first example the second can be skimmed, but if it is not clear from the first the second can become crucial. Over 70% of the real examples come from Canadian data. On the other hand I have taken examples from Belgium, England, Mexico, and the US where these have seemed clearer, more memorable, or simply more likely to interest my students than Canadian examples of which I am aware.

The book has been strengthened considerably by suggestions from reviewers, who typically were in broad agreement on what could helpfully be done. On one point, though, the best location for the chapters on statistical inference, they were divided. I have placed these chapters in accordance with majority preference, after the material on describing a single variable. However, having presented statistical inference at a later point in my own classes, and understanding well the advantages of doing so, I will point out that if two small sub-sections which rely on the idea of a sampling distribution are omitted or postponed, or if students remember the most basic ideas about inference, then Part Two on statistical inference can be moved without difficulty to after the section on measures of association.

Having profited from numerous suggestions from reviewers of the initial draft, I look forward to further suggestions from users of the book. Any received will be considered carefully.

Acknowledgements

If authors mention only those closest to the creation of their books, acknowledgements can be brief. If all those who directly or indirectly contributed are mentioned, the list can scarcely be completed. I will not attempt such a list, but I must mention those to whom most is owed.

I owe William and Rosalie Arnold a very great deal for encouraging their bookish son, at a time and place in which bookishness was not greatly in fashion. I owe them a great deal as well for taking out loans to put me through my first degree, at a time when student loans were unavailable, so that I could graduate without my own debts.

At the University of Saskatchewan, I was introduced to sociology by a great teacher, Edward Abramson, who encouraged me to take courses in both applied and theoretical statistics through the Department of Mathematics. Neither he nor I, nor my teachers Abraham Kaller and Nicholas Sklov could have known they were preparing me for much of the work of my adult life.

At the Social Planning and Research Council in Hamilton, where I had been hired by Harry Penny, soon to be the Director of the School of Social Work at McMaster, I was fortunate to be principal investigator for three studies that resulted in book-length manuscripts, each based on quantitative data. Reports from each were reviewed by an advisory committee of academics and social service administrators. For most of my time there, the chair of the committee was Frank Jones of the McMaster Sociology department, later to be a member of my doctoral committee. Some examples in this text stem from work done at the Council.

Michael Wheeler, at that time in the School of Social Work at McMaster, and Doris Guyatt, of the Ontario Ministry of Community and Social Services, then provided the opportunity to serve as principal investigator in what turned out to be a three-wave longitudinal study of the experience of parents with custody of children after marital separation. Some examples in the text stem from this work.

During my years in applied research I always enjoyed opportunities to teach. With that in mind, I went to McMaster for doctoral work. There I had the opportunity to do a specialized comprehensive in quantitative methods, which I chose, in part, to allow me to make contributions to joint projects, from which I have since learned much. At McMaster I also learned a great deal about how to teach statistics to students of sociology by watching Alfred Hunter, Charles Jones, and Peter Pineo at work. Peter Pineo, my dissertation supervisor, in his self-effacing way, was a model of careful analytic practice. No one's work was more certain to replicate well. The exposition of ANOAS models for my dissertation helped to prepare me to write the chapter on log-linear modelling available from the textbook website.

At Queen's I had the opportunity to teach methods and statistics at both the undergraduate and the Master's levels. After two years, though, my teaching had to be reduced because I had become project methodologist for the evaluation of Better Beginnings, Better Futures, a provincially sponsored program designed to improve the life chances of children and families in disadvantaged neighbourhoods. Numerous examples in this volume are based on the Better Beginnings data set.

I am grateful to Vince Sacco for introducing me to the project, which, with its multi-site longitudinal design, provided ongoing challenges. I am also grateful to the funders, Ray Peters, Kelly Petrunka, and the other members of the project Core Team for providing a context in which my role in a highly complex evaluation could be carried out.

Having moved to Windsor, I have had the opportunity teach statistics and/or methods every year at the undergraduate level, and often at the M.A. or Ph.D. level. This experience has helped greatly in thinking through what can and should be presented, and how, in a textbook of the kind you have before you.

The final form of the manuscript owes a great deal to the staff of Oxford University Press Canada.

Thank you to the reviewers of the draft manuscript: Tom Buchanan, Kenneth MacKenzie, Jenny Godley, and those who chose to remain anonymous. Of course I must accept responsibility for such failings as may remain.

S.D.G.
Robert Arnold,
September, 2014

Part I
DESCRIPTIVE STATISTICS FOR ONE VARIABLE

Descriptive statistics for a single variable often provide key pieces of information. The mid-term average gives an indication of how well a class is doing. Unemployment rates provide an indication of where an economy is moving. Average incomes for immigrant groups inform us about how well they are integrating into the Canadian economy.

Besides providing key information, descriptive statistics may tell us how far we can rely on other data. Survey researchers report characteristics of their samples—age, sex, income, and the like—so readers can decide whether the samples are biased. For studies of specific populations, statistics may be needed to show that they have been reached. For example, I once looked at contraceptive practices in low-income areas, and had to show that the people interviewed were often of lower socio-economic status.

Descriptive statistics can also tell us what analyses can be done. If, for example, we want to know how cultural or racial groups are faring in the economy, we have to see how many cases we have from which groups. Or if we hypothesize that the effect of income on health will be most pronounced at the lower extremes of income, we will need to see how many cases we have there.

It is for good reason, then, that this text begins with descriptive statistics for single variables. Since the ones we use depend heavily on levels of measurement, we shall begin with them.

Levels of Measurement

Learning Objectives

In this chapter, you will learn

- a widely known way of classifying variables, which helps us to decide which statistics can be used with them;

- some limitations of this classification;

- two other common ways to classify variables; and

- typical ways of graphing for variables of specific types.

The Stevens Classification

The most widely used classification of levels of measurement is that proposed by Stevens (1946). We shall see how he distinguished among four levels and look at some of their limitations in practice. Two alternative classifications will be considered, and graphs appropriate to the second of these will be illustrated.

For Stevens (1946), measurement requires

a) that we be able to categorize each observation, and

b) that we be able to place each in only one category.

For example, if we ask people their country of birth, we should be able to place each person in a category: Canada, Italy, Somalia, and so on, and each person should fit in only one category. To put the requirements a bit more technically, the categories must be exhaustive (so all cases can be placed) and mutually exclusive (so no case can go into more than one).

When these standards are met, measurement can take place at four levels:

nominal;

ordinal;

interval; and

ratio.

Nominal Measures

What distinguishes nominal measurement is that any numbers assigned to the categories are arbitrary. We might set up a code scheme beginning with

1 - Canada;

2 - Italy; and

3 - Somalia,

but we could equally well have set it up as

1 - Somalia;

2 - Italy; and

3 - Canada.

In the social sciences we have many nominal measures, with numbers assigned arbitrarily. We very often see

1 - male; and

2 - female,

but we also often see

0 - male; and

1 - female.

Again, we may see main activity coded

1 - employment;

2 - seeking work;

3 - schooling;

4 - homemaking;

5 - retirement; and

6 - other,

but the numbers could easily be switched around.

Because numbers are assigned arbitrarily in nominal measurement, it makes no sense to do arithmetic with them.

Ordinal Measures

When we can meaningfully order, or rank, the categories, we can assign numbers reflecting their order. Suppose we ask for responses to "I am enjoying my time at university." Answers can be put in an order reflecting degree of agreement—for example,

1 - strongly agree;

2 - agree;

3 - neutral;

4 - disagree; and

5 - strongly disagree.

The interval and ratio levels also require that categories be ordered. What distinguishes the ordinal level is that we cannot readily say that the distance between any two categories is the same as that between any two others. In the example, we cannot readily say whether the difference between "strongly agree" and "agree" is the same as that between "neutral" and "disagree" or between any other two adjacent categories.[1]

The same applies for many other sociological variables—for example, highest level of formal education, which might be coded

0 - secondary incomplete;

1 - secondary complete;

2 - some post-secondary, no degree; and

3 - university degree(s).

Any category further along is more advanced than those before it, but the distances among them are unclear.

Because we do not know the precise distance between categories, we are in an awkward position to do arithmetic. With full certainty, we can only treat one category as above or below another. However, as we shall see, those who believe that distances between categories are approximately the same often use addition and subtraction. Their reasons will be illustrated after we look at Stevens' other two levels of measurement.

Interval Measures

At the interval level, as in all successful measurement, categories are exhaustive and mutually exclusive. As at the ordinal level, they are clearly ordered. In interval measurement, though, the distance between adjacent categories is the same wherever we are along the scale. A stock example is temperature. The distance between 5 degrees and 10 is the same as the difference between 15 and 20.

Because the distances are precise, and adjacent categories are equidistant, we can do addition and subtraction straightforwardly. If we take 10 degrees − 5 degrees = 5 degrees, the result means the same as if we had taken 20 − 15.

We still have a problem with multiplication and division, though. If the temperature were 5 degrees, and someone suggested that at 10 degrees it would be twice as warm, we would be puzzled by the claim. The reason is that an interval measure lacks a meaningful 0. The mere existence of the Celsius and Fahrenheit scales, with their differing 0 points, underscores the reality that 0 does not represent the full absence of temperature. That being so, it makes no sense to say that 10 degrees is twice as hot as 5, or half as hot as 20.

Ratio Measures

As at the earlier levels, categories are exhaustive and mutually exclusive. As at the ordinal and interval levels, categories are ordered, and as was true at the interval, adjacent categories are equidistant. The distinctive feature of the ratio level is a meaningful zero point. Examples include population counts, grades on examinations, incomes, and years of age or of education completed.

Because there is a meaningful zero, not only addition and subtraction, but also multiplication and division make sense. If we multiply someone's income by two, we get the income of someone who, in these terms, is doing twice as well. If a mother is 52 and her daughter 26, it is meaningful to say the daughter is half her mother's age.

Dichotomies

Dichotomies (variables which take on only two values) are a special case. They may be treated as lying at any one of the four levels. Take, for example, a variable called UDEGREE, coded

0 = has no university degree; and

1 = has a degree.

If the order of the categories is of no concern—we just want to distinguish two groups— there is no problem in treating this variable as nominal. It can also be seen as ordinal, since having a degree means having more advanced education than being without one. Further, since there is only one distance between categories, there need be no concern about whether distances are constant, so this variable can be treated as interval without a problem. We can also think of the first category as representing a meaningful zero: you have a degree or you do not, so if sheer presence of the degree is what matters, the variable can be treated as ratio.

Table 1.1: Distinguishing Characteristics of Stevens' Levels

	Nominal	Ordinal	Interval	Ratio
Categories				
Ordered		yes	yes	yes
Equidistant			yes	yes
Scored from true 0				yes
Arithmetic Possible Using Categories				
Distinction of categories	yes	yes	yes	yes
< or > comparison		yes	yes	yes
Addition and subtraction		yes?[2]	yes	yes
Multiplication and division				yes

Figure 1.1: Questions to Determine Levels of Mveasurement in Stevens' System

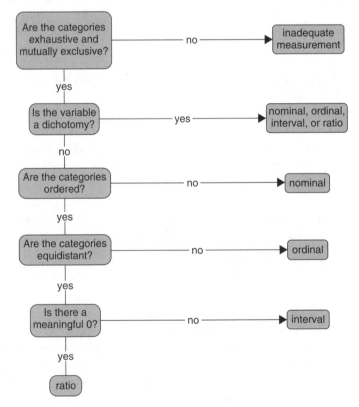

Treating Ordinal Variables as Interval

In the social sciences we have few unambiguously interval variables. That is one reason why the stock example, temperature, is not drawn from the social sciences. We use many variables which are ordinal in Stevens' terms, though, and often treat them as interval, adding and subtracting their scores. The vast bulk of quantitative social scientists do this, at least sometimes, so we need to see why they find level of measurement an insufficient guide to practice. We shall look at construction of scales, and one common situation in which measures designed for ordinal variables do not work well.

Scale Construction

We often cannot believe that a single question will properly measure an attitude or internal state. For example, researchers have put a good deal of work into assessing the social causes of depression. It is usually assessed in terms of symptoms, but no one of these is sufficient to identify the condition, so several must be considered. Respondents may be asked to say whether they feel sad, feel like crying, feel tired, and so on. Their answers may be put into a category scheme such as

0 - rarely or never;

1 - a little of the time;

2 - some of the time; or

3 - most of the time.

The categories are ordered, but the distances between them are imprecise.

Two rationales can be offered for adding up the scores. The first is that our question is not what accounts for the prevalence of individual symptoms, but rather what accounts for the prevalence of the depressive syndrome, which can only be assessed by multiple items. To combine them, and thus get an answer to our question, we have to add the scores. We get an approximation to what we would get if we knew the precise distances between categories, but this is better than no measure at all.

The second rationale is that the sum of scores across multiple items is likely to be much more reliable than that for a single item. Answers to single questions about attitudes or internal states often change a good bit over short periods of time. If the fluctuations are random, then those for one question will tend to cancel out those for another. The non-random element in the answers will cumulate from question to question. The result is that the sum of scores to a set of items is typically more reliable than the individual item scores. The increased reliability of the total may well outweigh the imprecision involved with adding up the scores.

The case made for summing scores across items may, in a given instance, rely on both rationales.

Using Statistics That Assume Equal Intervals

Apart from scale construction, researchers often treat ordinal variables as interval when no method not assuming equal intervals is workable. Again, an approximation is often seen as preferable to the best procedure available without treating the data as interval.

Consider taking the mean for a variable coded

1 - strongly agree;

2 - agree;

3 - disagree; and

4 - strongly disagree.

To get a mean, we have to sum the scores. To get the obvious alternative, a median, we just have to order the cases and find the value that splits the lower half from the upper. Since no addition is required, the median poses no logical problem. The catch is that changes in the median may not track changes in the data well, as illustrated in Figure 1.2.

Figure 1.2: Shifts in the Mean and Median

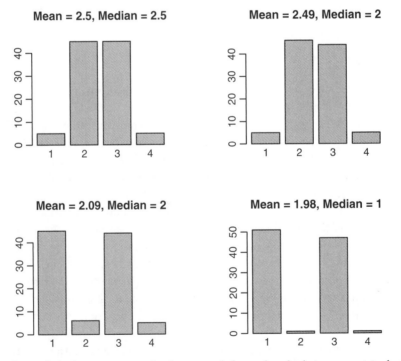

The graphs are based on 100 cases. In the upper left graph, which is symmetric, both the mean and median are 2.5. On the upper right, we see what happens when two cases move from category 3 to category 2. The median drops abruptly to 2, while the mean shifts subtly to 2.49, reflecting the minor change that has occurred. On the lower left, 43 cases move from category 2 to category 1, but the median shifts not at all. The mean drops to 2.09, reflecting the sizable change in the data. Finally, on the lower right, 3 cases have moved into category 1, creating an abrupt drop in the median from 2 to 1. At the same time, mean slides from 2.09 to 1.98, responding in a moderate way to a modest change in the data.

The median, responding only to what happens at the exact centre of the data, moves not at all in response to major changes elsewhere, but may shift sharply in response to movement of a few cases near the centre. The mean responds to changes anywhere and changes in proportion to the movement in the data. Many prefer the mean over the median for these reasons. It provides an approximation to what could be obtained if distances were precisely known, but the alternative is a very unsubtle measure of change. Another situation in which measures designed for ordinal variables are problematic will be seen when we consider the standard deviation.

Of course, relying on approximation requires that distances between scores be roughly alike. Some suggest that many response patterns, including the Strongly Agree to Strongly Disagree pattern we have looked at, are somewhere between ordinal and interval; we don't know the precise distances between categories, but they are of roughly the same order. If the distances between categories do not vary wildly, why be forced into working as if they do? Stevens himself (1951:26) acknowledged the practical value of treating some ordinal variables as interval, noting that "in numerous instances it leads to fruitful results." Tukey (1961:245–6) remarked,

> Results based on approximate foundations must be used with the underlying approximation in mind. . . . But what knowledge is not ultimately based on some approximation? And what progress has been made, except with the use of such knowledge?

Another issue in working with Stevens' system is how to deal with variables that have some characteristics from one level and some from another. Often some categories can be ordered, but others cannot. If we ask parents how far they expect their children to go with their schooling, we may get answers such as

complete high school;

complete community college;

complete university; and

it's up to them.

The same issue can come up when coding highest completed level of education, as in this classification from Statistics Canada:

- less than Grade 9

- Grade 9, secondary incomplete

- secondary complete

- trades certificate or diploma

- some college

- some university

- university non-degree diploma

- university degree

Between "secondary complete" and "university degree" the ordering is often unclear.

A more recently noted difficulty with the Stevens classification lies in "fuzzy sets," for which category membership is uncertain, particularly around the edges. Good work has been done with them, but their use is a more advanced topic than can be covered here. Since statistical applications of fuzzy sets arose after Stevens' time, we ought not to be too critical about their omission from his scheme, but we need to be aware that what in his system would be a failure in measurement can sometimes lead to useful analyses.

Two Related Classifications

Two other classifications of variables are often seen. One, widely used in the discipline of statistics, distinguishes "qualitative" from "quantitative" variables. The difference lies between those with which arithmetic can straightforwardly be done and those with which it cannot. Thus qualitative variables are those which, under the Stevens classification, are labelled nominal or ordinal. Quantitative variables include those at the interval and ratio levels.

The second distinguishes among discrete and continuous variables, or adds a third category of discrete-continuous variables, as will be done here. A discrete variable can take on only specific values, distinct from other possibilities. Nominal variables are discrete, because cases must fall into specific categories, and these cannot be placed on a continuum. A strictly continuous variable is ordered, and there are no gaps in its distribution. A case may take on any value within the range, and possible scores may differ so little as to be virtually indistinguishable. Variables above the nominal level may be continuous, or may be discrete-continuous.

In the social sciences, we have many of the latter. These are well ordered, and their categories are often finely broken down, but there are gaps in their distributions. For example, population counts are limited to the integers. Again, income is typically reported in dollars, although it could be reported to two decimal digits. Even a variable like age, which could be reported to many decimal digits, is typically limited to the integers. These variables, and many others which are theoretically continuous, but for which our measurement is not continuous, count as discrete-continuous.

Graphing

Discrete, continuous, and discrete-continuous variables are typically graphed differently. Discrete variables are normally presented in bar charts, with spaces between the categories to emphasize their distinctness. Strictly continuous variables are not common in empirical social science, but theoretical variables may be continuous, and their distributions may be graphed as curves, or plotted in histograms. Discrete-continuous variables are normally presented in histograms.

The bar chart for main activity, shown in Figure 1.3, illustrates typical graphing for a discrete variable.

Figure 1.3: Main Activity of Canadians Aged 25–64, 2004

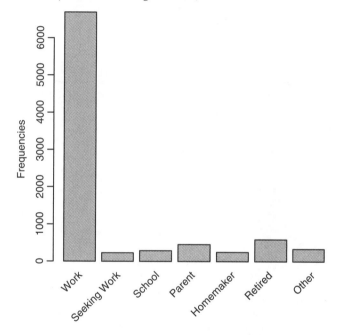

A clear example of a continuous variable is the classical standard normal distribution, typically graphed with a curve, and presented in Figure 1.4.

Number of evening visits to friends and relatives per month is a discrete-continuous variable, well suited to presentation in a histogram, as shown in Figure 1.5.

Figure 1.4: Standard Normal Curve

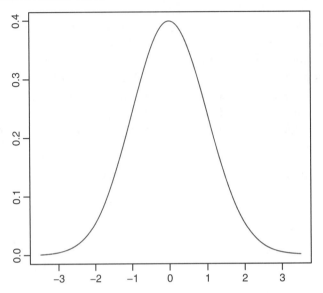

Figure 1.5: Evening Visits to Friends and Relatives

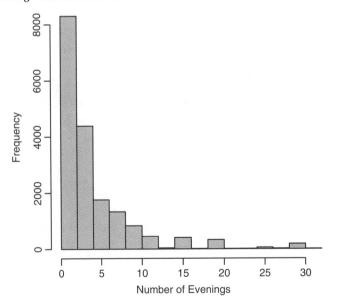

Summary

We have given most attention in this chapter to the classification of levels due to Stevens (nominal, ordinal, interval, and ratio), but we have also seen two other classifications. One, which uses the concepts of qualitative and quantitative, can be seen as a collapsed version of the Stevens classification, in which nominal and ordinal variables are labelled qualitative, and ordinal and interval variables are labelled quantitative. The other system, which uses the concepts of discrete, continuous, and discrete-continuous, is linked to our choices in graphing.

As well as considering concepts, we have seen some issues in their application. The issue of treating ordinal variables as interval will reappear in later chapters. In fact, concepts from this chapter will organize much of what appears in the several that follow. In Chapter 2, we will see that our choice among the mean, the median, and the mode must be affected by level of measurement (although other considerations will apply). In Chapter 3, we will see that our choice among measures of how widely dispersed our data is also requires consideration of level of measurements. The same will hold for measures to be discussed in other chapters.

The moral is that the material in this initial chapter is fundamental to social statistics, so that those who master it early will be advantaged in much of what lies ahead.

Review Questions on Levels of Measurement

1. List Stevens' four levels of measurement. Explain the characteristics of one of them.

2. What are the differences between a nominal and an interval variable (or between an ordinal and a ratio variable)?

3. According to Stevens, what kinds of arithmetic can we do with ordinal variables? With ratio variables?

4. What rationales are sometimes given for treating a set of questions on a particular topic—e.g., perceived level of social support—which are themselves ordinal, as if they were interval?

5. What are two situations in which social scientists often treat ordinal variables as interval?

6. What class of variables can be treated as lying at any one of Stevens' levels? Why?

7. Why has the mean often been considered preferable to the median for some variables that, strictly speaking, are ordinal?

8. What are "qualitative" and "quantitative" variables?

9. What is a discrete variable? A continuous variable? A discrete-continuous variable?

10. How are discrete, continuous, and discrete-continuous variables typically graphed?

11. What is the difference between a bar chart and a histogram? What sort of variables are typically placed in bar charts?

12. Suppose someone was not sure what level of measurement a variable had reached (in the Stevens system). What set of questions should make it possible to work this out?

Notes

1. The distances between categories can be estimated through methods grouped under the heading of Item Response Theory, but these methods are topics for more advanced courses.

2. Researchers who think of items as approximately interval often add or subtract item scores, and use such items or scales created from them in analyses developed for interval data. Since this is widespread, but not accepted by strict believers in Stevens' system, a question mark has been placed in the table.

Measures of Central Tendency

Learning Objectives

In this chapter, you will learn

- how to choose among common "averages"—the mean, median, and mode—based on levels of measurement;

- other important considerations in choosing among them;

- what we can do when they might be misleading;

- some advantages of the median and mean that are not widely recognized; and

- relationships among the three basic measures.

The most widely known descriptive statistics are measures of central tendency. Some of them—the mean, median, and mode—are often introduced by the middle elementary grades. Here we need to consider

- how they are linked to levels of measurement,

- how we should choose among them, and

- what we should do if they are liable to be misleading.

We must also be clear on the varied senses in which they tell us about the "centre" of the data.

Definitions and Notation

First, a very brief review of definitions and notation. The mode, often denoted "Mo," is simply the most frequently observed category. The median, sometimes denoted "Md," is the value that, after the observations have been placed in order, divides the lower half from the upper half.

The median is often referred to using percentile notation. Percentiles are the points that divide an ordered set of cases into 100 categories of equal size; for example, the tenth percentile (P_{10}) marks off the lowest 10% from the upper 90%. The fiftieth, which divides the lower from the upper half, is equal to the median, so we can refer to the median as P_{50}.

The mean is obtained by summing the scores for all cases, then dividing by the number of observations. In standard notation, the mean is

$$\Sigma\, x_i\, /\, N \, ,$$

where the uppercase Greek sigma, the summation sign, tells us to add up the values referred to on its right. The x_i are the individual values of x. After summing them, we divide by the number of cases we have, denoted here by N.

The mean is often denoted by a bar, or macron, placed over the name for a variable, as in $\bar{x}$, the mean of the variable x. (In speech we refer to it as "x-bar.")

Links to Levels of Measurement

Measures of central tendency are linked to levels of measurement in what at first seems a straightforward way. If we have nominal data, we cannot arrange cases in order, so we cannot calculate the median. We cannot do addition or subtraction either, so the mean, which relies on addition, is unavailable. The only measure we have is the mode. It gives us only the identity of the largest category, but that is all that can be had.

With ordinal data, we can use the mode if we wish to single out the largest category, but the mode does not take advantage of our ability to order the categories. To calculate a mean requires addition, which requires justification if we do not know the exact distances between categories. The median, which uses our knowledge of the order of the categories, without assuming well-defined intervals,[1] is often the first suggestion for ordinal data.

With interval or ratio data, we can report on the largest category, or on the point which divides the cases into upper and lower halves. Since the distances between categories are well defined, we can also add scores to calculate the mean. Because it takes advantage of our precise knowledge of distances between observations, it is typically the measure first suggested for interval or ratio data.

Default options are sometimes set out on the basis of level of measurement: use the mode for nominal data, the median for ordinal, and the mean for interval or ratio data. It is typically added, though, that it is logically acceptable to use a measure suited to a level of measurement below that of the data. In fact, if we have interval or ratio data, and we have reason to point out the value which divides our observations into upper and lower halves, the median may be preferable to the mean. We may also prefer the median when it better represents the bulk of our cases. Consider Figure 2.1, which shows grades in an introductory sociology class.

Figure 2.1: Histogram of Grades in an Introductory Course

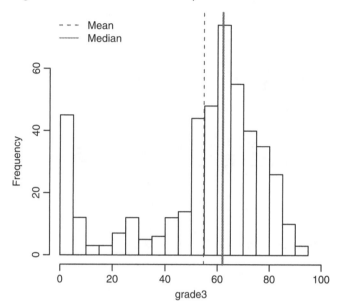

The mean, at just under 55, is noticeably below the median, at 62. The median is in the same category as the mode. The mean is pulled down by a subgroup of students who did little work, and whose final grades were under 10. The median better reflects the performance of the great majority of the class.

In fact, if we include only those who attempted 90% of the work—that is, who wrote exams and handed in papers that together were worth 90% of the final grade—there is little difference between the median and the mean. Perhaps in this case it would be better to report results separately for those who did the bulk of the work and those who did not, but if we have only the grades at our disposal, the median represents the bulk of the students better.

In other situations where we have ratio data, the mode draws attention to a key point and there is no reason not to report it. Consider Figure 2.2.

We could report either the mean (17.65 doctors per 10,000 population) or the median (16), but there the mode also points to something important. Of the 147 countries, 47, or almost a third, have fewer than 5 doctors per 10,000 population. To simply report the mean or median draws no attention to this important shortage.

For another example, consider the hours of work shown in Figure 2.3.

Figure 2.2: Doctors per 10,000 Population for 147 Countries, 2010

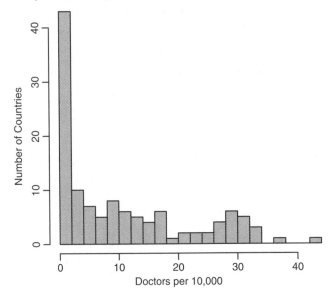

Figure 2.3: Weekly Hours of Paid Work, Canada, 2008

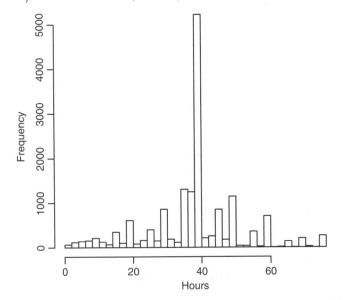

We have ratio-level data, so there is no logical problem in using either the mean or the median, but there is a standout category: well over three times more people work 39–40

hours than fall into any other range. Not to mention this modal category, and indicate how many fell into it, would be to miss a central feature of work in these data.

From what we have seen, selecting a measure requires us to decide

1.   which measure is immediately suggested by the level of measurement;

2.   which others are logically appropriate; and

3.   how helpful are the messages conveyed by each.

However, two further issues remain:

4.   might a measure be unstable, and hence potentially misleading?; and, if so,

5.   what alternative, or modification, will reduce the risk of being misled?

Potentially Unstable Results

If we gather several sets of data from the same population, we can expect measures of central tendency to vary at least a little from one to the next. There may be slight differences in the people from whom we try to gather data; somewhat different people may be available on one occasion than on another; people may respond a bit differently to the interviewers; and interviewers may ask questions in slightly different ways. If, for reasons of this kind, results differ from one sample to the next in an important way, our measurement is said to be unstable.

Instability of the Mode

Each measure of central tendency can be unstable under the wrong circumstances. The mode becomes unstable when two or more categories are about equally common. In the bar chart in Figure 2.4, a few more cases would shift the mode from College to High School, and a few more would shift it to Some College.

The obvious solution when categories have quite similar frequencies is to report the largest, but to mention that it has close competition and to identify its competitors.

Instability of the Median

As we have seen above, the median can change very quickly when the cases at the very centre of a distribution move from one category to another. As a consequence, the median becomes unstable when cases are sparse near the centre of the distribution, as in the histogram in Figure 2.5. If there are large distances between cases near the centre, a few more cases on one end or the other can shift the median substantially. This problem is most apt to arise with small samples. The informally labelled "bathtub" distribution illustrated in Figure 2.5 may be unusually vulnerable to it.

Figure 2.4: Husband's Education

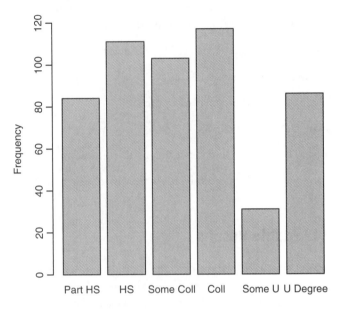

Figure 2.5: Histogram of a "Bathtub" Distribution

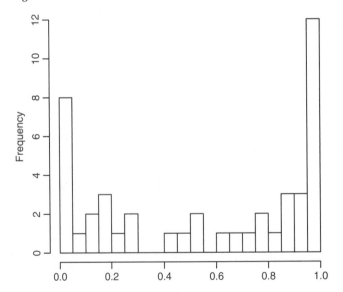

When the median appears unstable, a commonly suggested strategy is to use more information. The median, after all, is based on the value of a single observation (or the midpoint between two observations) in the middle of the distribution. Using the values of other observations ought to stabilize it. One possibility is to use a broadened median, taking the mean of the median itself and one or two observations on each side. Another is to use the values of percentiles near the centre. For example, as a substitute when the median is unstable, we might take $(P_{40} + P_{50} + P_{60}) / 3$. Another suggestion is the mid-mean, that is, the mean of all observations in the central half of the distribution. Since the mid-mean uses more data than the other alternatives to the median, it is likely to be stabler.

Each of these solutions requires us to add up the values for cases—i.e., to treat the data as at the interval level. With ordinal data, we may well prefer the stable approximation obtained through this addition to an unstable median. If not, we will have to fall back to the mode or live with an unstable median.

Instability of the Mean

The mean becomes unstable in the presence of outlying cases, which, if extreme enough, can shift it greatly. Income distributions often include outliers, so that the mean is often an awkward measure of central tendency for them. Consider the distribution in Figure 2.6.

Figuve 2.6: Annual Household Income, Canada, 2007

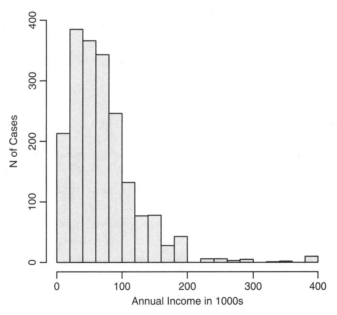

In the presence of outliers, two alternatives are widely used. The first is a trimmed mean, in which we remove the upper and lower N% of observations before recalculating the mean. The fraction removed varies, but often a trimmed mean is based on between 95% and 99% of the cases. (The mid-mean, mentioned above, is a very vigorously trimmed mean.) The second approach is to use a measure based on percentiles which is not sensitive to the values of extreme cases, typically the median.[2] Trimmed means use more of the data, but the median offers the useful interpretation that it divides the distribution into upper and lower halves.

For skewed distributions, trimmed means often come closer to the median as more cases are trimmed. For the income data here,

the median is 63,000; and

the mean is 75,231; while

a 99% trimmed mean is 73,169;

a 95% trimmed mean is 68,954; and

a mid-mean is 64,495,

much closer to the median of 63,000 than the less vigorously trimmed alternatives.

The median is often used with income, because extremes can make a major difference, because cases near the centre are typically abundant, and because the measure is readily interpreted.

Both the median and trimmed means are "resistant" in the sense that they are not affected by, or resist the influence of, extreme cases. Another way to "resistify" the mean, of which you should be aware, is to assign cases lower weights as we move into the tails of the distribution. A fourth method, known as "Winsorizing" (after Charles Winsor), is to recode the values of extreme cases so they equal others a bit closer to the centre, perhaps to equal the 1st or 99th percentile or a value a little less extreme.

Measures of Central Tendency as Averages

The mode is an "average" in the sense that it is the most typical value in a distribution. Sometimes, though, it is not very typical: if we have multiple categories, the mode may include only a modest fraction of the cases. That is why, when reporting a mode, the number or proportion of cases in it should be given.

The median is a "positional average": when the cases are ordered, it lies precisely in the middle. A less obvious advantage is that the sum of absolute deviations from the median is at a minimum. A straightforward illustration uses three values:

3, 7, 17.

Their median lies at the second value, 7.

We calculate the sum of absolute deviations straightforwardly:

$(x_i - 7)$	$abs(x_i - 7)$
$3 - 7 = -4$	4
$7 - 7 = 0$	0
$17 - 7 = 10$	10
	$\overline{14}$

To see how the sum changes if we start from somewhere other than the median, let us begin from the mean, which is 9. The sum of absolute deviations is higher by 2 than the sum of deviations from the median.

$(x_i - 9)$	$abs(x_i - 9)$
$3 - 9 = -6$	6
$7 - 9 = -2$	2
$17 - 9 = 8$	8
	$\overline{16}$

No other value can give us a lower sum of absolute deviations than the median does. To illustrate, Figure 2.7 shows the sums for values in the range of the example data—that is, from 3 to 17. Treating each value as a central point, we see that the sum of the deviations drops consistently as we move toward the median from the left and rises as we move away to the right.

Figure 2.7: Sums of Absolute Deviations from Points in the Range 3–17

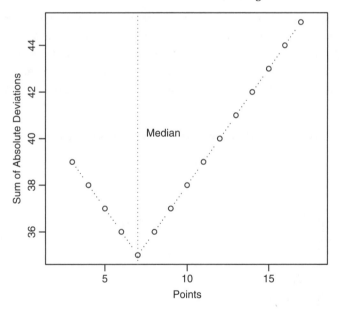

The mean too offers advantages that are not widely appreciated. It is the central point in that the sum of positive departures from it, where $x_i > \bar{x}$, equals the sum of negative departures, where $x_i < \bar{x}$. Another way to put this is to say that, as an average, it neither over- nor underestimates.

For this to hold, the sum of all departures from $\bar{x}$ must equal zero. Algebraically, it must be true that

$$\Sigma (x_i - \bar{x}) = 0$$

To show that this holds, we rewrite the equation as

$$\Sigma\,(x_i - \bar{x}) = 0$$

Then $\Sigma\,x_i - \Sigma\bar{x} = 0$.

Since $\bar{x}$ is a constant, its sum, across N cases, is just $N\bar{x}$, so that

$$\Sigma\,x_i - \Sigma\,\bar{x} = \Sigma\,x_i - N\bar{x}$$

The mean itself is just the sum of individual values of a variable, divided by N. So, for the variable x,

$$\bar{x} = \Sigma\,x_i\,/\,N$$

Substituting $\Sigma\,x_i\,/\,N$ for $\bar{x}$, we get

$$\Sigma\,x_i - N\bar{x} = \Sigma\,x_i - N(\Sigma\,x_i\,/\,N)$$

Cancelling the Ns in the right-hand term, we have

$$\Sigma\,x_i - \Sigma\,x_i = 0 \text{ , as required.}$$

The mean is also central in that the sum of squared deviations from it is lower than the sum from any other point. A proof requiring only straightforward algebra is appended.

The graph in Figure 2.8 illustrates. Using the data values 1 through 9, we obtain a mean of 5. The sums of squared deviations obtained by treating each value from 1 through 9 as the central point are plotted in the graph. These decline as we approach the mean from the left, and rise as we move away to the right.

Figure 2.8: Sums of Squared Deviations from Central Points in the Range 1–9

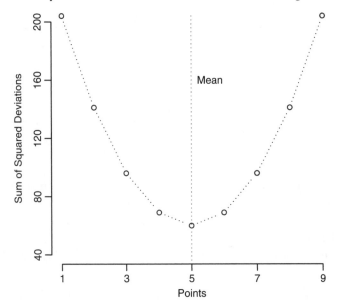

Relations among the Mode, Median, and Mean

It is often pointed out that for single-peaked, skewed, and continuous distributions, the mean lies farther into the long tail than the median, while the mode lies at the peak. This point is illustrated in Figure 2.9.

Figure 2.9: Mode, Median, and Mean for a Continuous Right-Skewed Distribution

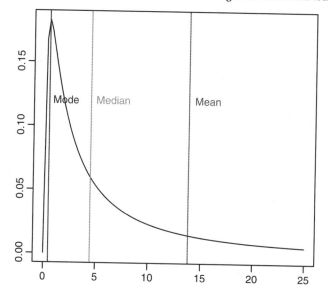

The same relationships among measures of central tendency hold for many distributions, but not for all, as can be easily shown with discrete distributions. Suppose we have the following seven values, arranged in order:

   1, 2, 3, 4, 5, 5, 22

The median is the fourth of the seven values, or 4. Only one value, 5, occurs more than once, so 5 is the mode. The mean is

$$1 + 2 + 3 + 4 + 5 + 5 + 22 = 42 / 7 = 6$$

Here, although the mean (6) is the highest, the order of the mode (5) and the median (4) is reversed.

   It is also easy to show that the mean can lie between the other two. Suppose we have these 11 values:

   0, 1, 4, 4, 7, 9, 10, 11, 13, 14, 15

The median is the sixth value, or 9. The mode lies at 4, since 4 is the only value that occurs more than once. The mean is

$$0 + 1 + 4 + 4 + 7 + 9 + 10 + 11 + 13 + 14 + 15 = 88 / 11 = 8 ,$$

so we have Mode < Mean < Median.

The point is simple: for single-peaked, skewed, and continuous distributions, the mean will lie farthest into the long tail, followed by the median, and the mode will lie in neither tail. While the ordering of the measures is often the same for distributions of other kinds, it does not have to be so.

Summary

Although the mean, median, and mode are widely familiar—in part because they are very useful—the considerations we ought to apply in using them are much less familiar. We have seen that levels of measurement offer initial guidelines for choosing among them, but that sometimes more than one of them conveys valuable information, so that we should employ more than one. We have seen as well that data that cannot be well analyzed if treated as ordinal are often treated as interval. Then too, each of these measures can be unstable under the wrong circumstances, and we must be ready to deal with instability.

Our reasons for believing that the median and mean are good measures of central tendency are often unfamiliar as well. We have seen that the median is the point from which the sum of absolute deviations is minimized. The mean is the point at which the sum of positive deviations and the sum of negative deviations balance. In this sense, it neither over- nor underestimates, on balance. It is also the point for which the sum of squared deviations is minimized. This fact will come up again when we look at regression.

Review Questions on Central Tendency

1. What rules of thumb are sometimes offered as to how levels of measurement should be associated with measures of central tendency?

2. What additional considerations must be borne in mind?

3. What are three situations in which we might want to use a mode, and one in which it could be misleading?

4. Define the median. What are three situations in which we might want to use it?

5. What is one situation in which a median could be misleading? What might we do in such a situation?

6. What is one situation in which we might want to use a mean, and one situation in which it could be misleading?

7. Suppose that we have ratio data, e.g., earned income, but the sample is small and there is reason to think that the mean is unstable. What are some alternatives?

8. Why is the sample mean often replaced by, or supplemented by, the median for income distributions?

9. For skewed distributions, what often happens as we trim more cases from the ends of the distribution?

10. In what sense are the mode, median, and mean averages?

11. In what sense is the median the point closest to the observed data?

12. In what sense does the mean lie in the centre of a distribution?

13. What is another technical merit of the mean as a measure of central tendency?

14. In relations to one another, where do the mode, median, and mean lie if we have a continuous, single-peaked, and skewed distribution?

Notes

1. If the median lies between categories, we may place it halfway between them. Doing so is a convention designed to yield a reasonable figure, and does not imply that we know the precise distance between the two categories.

2. We might, as another option, choose, e.g., $(P_{40} + P_{50} + P_{60}) / 3$.

3

Measures of Dispersion

Learning Objectives

In this chapter, you will learn

- commonly employed measures of dispersion;
- how level of measurement affects our choice among them;
- limitations of each measure; and
- what we can do when they are liable to be misleading.

In the social sciences, when we see a measure of central tendency a measure of dispersion is usually not far away, because the spread of cases can provide very important information. Consider what may happen when we survey opinions on issues of the day. If someone were to ask whether our economic situation requires higher taxes on upper-income Canadians, and to use the responses "Strongly Agree" through "Strongly Disagree," the mean would probably be reported. By itself, though, it could tell us nothing about consensus or polarization. The graphs in Figure 3.1 show response distributions with the same mean, with the response categories scored from 1 to 5.

The first distribution might suggest that a well-considered package of increased taxes on the well-to-do could win majority support. The second might suggest that there would be much support, but also intense resistance from a large minority. The third might suggest that broad agreement could be reached on a moderate package of changes, but that without more strongly supportive voters there might be too little public interest to provide impetus for action.

Figure 3.1: Distributions with the Same Mean

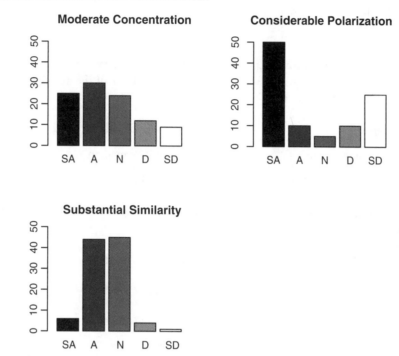

Because we often need to know about variation, much effort has gone into developing measures of dispersion. Like those for central tendency, they must be selected with an awareness of the levels of measurement for which they make sense. In turn, we will consider measures designed primarily for nominal data, for ordinal data, and for interval or ratio data.

Measures for Nominal Variables

The first of the measures for nominal variables, the Index of Diversity, tells us how likely it is that two cases, drawn at random, will come from different categories.

The Index of Diversity

Social scientists often consider forms of diversity. For example, we may wish to examine the extent of cultural diversity of cities, changes in the diversity of occupations held by women, or the progress of a firm in recruiting from different cultural groups.

The Index of Diversity is based on the probability that two people, chosen at random, will come from different categories. If there were no diversity, with everyone in the same category, then the probability would be 0. As the number of categories increases and their size decreases, the probability rises.

Since people are in the same category or they are not, we can write the Index of Diversity as 1.0 minus the probability that cases will match:

$$1.0 - p(\text{matching}),$$

where by "matching" we mean that people are in the same category.

As a first illustration, consider the cultural identities of one of the author's classes, in which there were eight students.

- American   1
- Canadian (anglophone)   4
- Italo-Canadian   1
- Indo-Canadian   1
- Trinidadian   1

For a case drawn at random, the probability that a given category would come up would be as follows:

- American $\frac{1}{8}$
- Canadian (anglophone) $\frac{4}{8}$
- Italo-Canadian $\frac{1}{8}$
- Indo-Canadian $\frac{1}{8}$
- Trinidadian $\frac{1}{8}$

We can write these probabilities as n_j / N, where n_j is the number in a category and N is the total number of cases.

After drawing a case, there would be one less in its category, and one less in the sample. If any one of the categories came up, the probability of drawing a match would be as follows:

- American $\frac{0}{7}$
- Canadian (anglophone) $\frac{3}{7}$
- Italo-Canadian $\frac{0}{7}$
- Indo-Canadian $\frac{0}{7}$
- Trinidadian $\frac{0}{7}$

We can write these probabilities as $(n_j - 1) / (N - 1)$.

The probability of drawing a match for a specific category is given by the probability of drawing that category initially times the probability of getting it again on a second draw. We can write this as

$$\frac{(n_j)}{(N)} \times \frac{(n_j - 1)}{(N - 1)} \quad \text{or} \quad \frac{(n_j)(n_j - 1)}{N(N - 1)} \qquad [1]$$

Using the formula on the left, we can calculate the following:

- for American:

$$\left[\frac{1}{8}\right]\left[\frac{0}{7}\right] = \frac{0}{56} = .000$$

- for Canadian:

$$\left[\frac{4}{8}\right]\left[\frac{3}{7}\right] = \frac{12}{56} = .214$$

- for Italo-Canadian:

$$\left[\frac{1}{8}\right]\left[\frac{0}{7}\right] = \frac{0}{56} = .000$$

- for Indo-Canadian:

$$\left[\frac{1}{8}\right]\left[\frac{0}{7}\right] = \frac{0}{56} = .000$$

- for Trinidadian:

$$\left[\frac{1}{8}\right]\left[\frac{0}{7}\right] = \frac{0}{56} = .000$$

To get an overall measure of the probability of getting a match, we sum across groups, obtaining

$$.000 + .214 + .000 + .000 + .000 = .214$$

To express what we have done in a formula, we simply add a summation sign to [1], and get

$$\frac{\Sigma\,(n_j)\,(n_j - 1)}{N(N-1)}$$

To this point, we have looked at the probability of a match, but for the Index of Diversity we need the probability of not getting a match, which as we have seen is $1.0 - p(\text{matching})$. Here, this is

$$1.0 - .214 = .786$$

From what we have seen, a formula for the Index of Diversity can be written

$$\mathbf{1.0 - p(matching)} =$$

$$1.0 - \frac{\Sigma\,(n_j)\,(n_j - 1)}{N(N-1)} =$$

$$\frac{N(N-1) - \Sigma\,(n_j)(n_j - 1)}{N(N-1)}$$

The numerator can be simplified. Removing brackets gives

$$N^2 - N - \Sigma\,(n_j)^2 + \Sigma\,n_j$$

The sum of the n_j is just N, so we can substitute, giving

$$N^2 - N - \Sigma\,(n_j)^2 + N$$

Then we can cancel the negative and positive Ns to get

$$N^2 - \Sigma (n_j)^2$$

Bringing back the denominator leads to:

$$\frac{N^2 - \Sigma (n_j)^2}{N(N - 1)}$$

[2]

After this simplification, the logic behind the formula is not obvious, but it is still the case that

the Index of Diversity provides the probability that two cases drawn at random will be from different categories.

We can use 2006 Census data to look at diversity in place of birth across Canadian cities, taking the following categories:

Canada	United States	Caribbean
Central America	South America	Europe
Africa	Asia	Other

We cheerfully allow software to work out the indices, shown in Table 3.1.

Table 3.1: Indices of Diversity for 17 Canadian Cities			
St John's	.072	Ottawa	.341
Moncton	.077	Edmonton	.342
Quebec	.082	Victoria	.356
Charlottetown	.099	Montreal	.380
Sherbrooke	.123	Windsor	.410
Sudbury	.133	Calgary	.418
Halifax	.159	Vancouver	.579
Saskatoon	.169	Toronto	.648
Winnipeg	.329		

By this criterion, Canadian cities vary in diversity by a factor of nine.

For another illustration, let us consider occupations of males and females. The 2006 Census categorizes them as shown in Table 3.2.

Plainly, males are more concentrated in some fields and females in others, but the overall level of diversity is not obvious for either group without a measuring instrument. Yet a clear comparison bears on the extent to which women, relative to men, have been restricted in occupational choice. The Index turns out to be somewhat higher for males, at .846, than for females, at .807.[1]

Table 3.2: Occupational Categories of Canadian Males and Females, 2006

	Males	Females
Management	11.6	7.5
Business, finance, administration	9.7	27.1
Natural sciences	9.7	3.0
Health	2.1	9.6
Social sciences	5.2	12.1
Culture and recreation	2.5	3.5
Sales and service	19.3	29.1
Trades, transportation	26.7	2.2
Primary industries	5.7	1.8
Processing, manufacturing, utilities	7.5	4.1
	100.00	100.00

How finely the categories are broken down affects the value of the index, because the fewer cases there are in a category, the harder it will be to find matches for them. Accordingly, the categories must be similar for index values to be compared.

Often statistics are normed to lie between 0 and 1, but this one can reach 0 only if we have as many categories as there are cases. The Index is interpretable nonetheless, and its meaning can be readily explained to non-statisticians. Those who prefer a measure to lie in the 0 to 1 range may be attracted to the Index of Qualitative Variation, which has the same numerator, but whose denominator is adjusted to keep values between 0 and 1.

The Index of Qualitative Variation

In short, the Index of Qualitative Variation (IQV) is a rescaled version of the Index of Diversity. The two have the same numerator, but the IQV's denominator is set up to ensure that the IQV will range from .00, when there is no variation (all cases are in one category), to 1.00, when there is maximum possible variation (that is, when cases are evenly dispersed across categories.)

Rationale of the IQV

As for the Index of Diversity, if cases are in the same category, they are said to match. If all are in the same category, all are matches, and a measure of variation should equal .00. To the extent that more cases are in one category than another, there is concentration. Thus, we have the minimum of concentration when each category contains the same number of cases. In this situation, the greatest possible number of unmatched pairs will be found.

The IQV is obtained from

$$\frac{\text{(Number of Unmatched Pairs)}}{\text{(Greatest Possible Number of Unmatched Pairs)}}$$

If there are no unmatched pairs (if all cases are in one category), the numerator will be 0, so the IQV will be .00. If the number of unmatched pairs is as great as possible, then the numerator will equal the denominator, so the IQV will be 1.00.

The numerator of the IQV is the same as that of the Index of Diversity, so we can borrow the formula

$$N^2 - \Sigma\, n_j^2 \qquad\qquad [1]$$

We just need to see how to get the greatest possible number of unmatched pairs for the denominator.

The highest number is obtained when there is no concentration, when each category contains the same number of cases. If there are k categories, each contains N/k cases.

The number of matched pairs in any one of them will be given by

$$n_j(n_j - 1) = (N\,/\,k)(N\,/\,k - 1) \qquad\qquad [2]$$

The number of matching pairs will be the same for each category, so the sum of matched pairs, across categories, will be k × [2], or

$$[k]\,(N\,/\,k)(N\,/\,k - 1)$$

Cancelling ks gives

$$N(N\,/\,k - 1)$$

To get the maximal number of unmatched pairs, we can subtract the maximal number of matched pairs, which we get from the formula just above, from the total number of pairs, $N(N - 1)$. Thus the maximal number of unmatched pairs is

$$N(N - 1) - (N)(N\,/\,k - 1)$$

A little algebra gives us a simpler formula. Removing brackets yields

$$N^2 - N - N^2/\,k + N$$

Cancelling −ve and +ve Ns leads to

$$N^2 - N^2/\,k \qquad\qquad [3]$$

Taking the number of unmatched pairs from [1] and the greatest possible number from [3] yields

$$IQV = \frac{N^2 - \Sigma n_j^2}{N^2 - N^2/\,k} \text{ , the standard formula.}$$

Suppose we want the IQV for the post-graduation plans of a fourth-year class of 10 students, whose plans as are shown below.

N = 10 (the number of students)

$n_1 = 3$ (professional school) $n_4 = 1$ (travel)

$n_2 = 2$ (MA studies) $n_5 = 1$ (undecided)

$n_3 = 3$ (employment)

k = 5 (the number of categories)

Substituting in $\dfrac{N^2 - \Sigma n_j^2}{N^2 - N^2 / k}$,

$$IQV = \frac{(10)^2 - \Sigma (3^2 + 2^2 + 3^2 + 1^2 + 1^2)}{(10)^2 - (10^2)/5}$$

$$= \frac{100 - (9 + 4 + 9 + 1 + 1)}{100 - 100/5}$$

$$= \frac{100 - 24}{100 - 20} = \frac{76}{80} = .95$$

The IQV tells us that the dispersion for this class is 95% of its possible maximum. That maximum would be reached if one student aiming at professional school and one aiming at employment were to shift, so there was an additional person intending to travel, and one more undecided. Then all categories would be equally frequent, at two students apiece, and dispersion could be no greater.

Comparing the IQV and the Index of Diversity

The IQV differs from the Index of Diversity only in the denominator. When samples are large, the two correlate very highly. Here we compare the results for 17 Canadian cities. The two measures are almost perfectly correlated, as shown in Table 3.3 and Figure 3.2.

A Conceptual Limitation of the IQV

A limitation of the IQV can best be seen by illustration. Suppose we have place of birth for two different samples, as shown on the left and the right.

Canada	2		Canada	2	Nigeria	2
India	2		Britain	2	Mexico	2
Nigeria	2		Germany	2	Somalia	2
United States	2		Hong Kong	2	United States	2

Table 3.3: Indices of Diversity and IQVs for 17 Canadian Cities

	Index of Diversity	IQV
St. John's	.072	.080
Moncton	.077	.087
Quebec	.082	.092
Charlottetown	.099	.111
Sherbrooke	.123	.138
Sudbury	.133	.149
Halifax	.159	.178
Saskatoon	.169	.190
Winnipeg	.329	.370
Ottawa	.341	.384
Edmonton	.342	.385
Victoria	.356	.400
Montreal	.380	.427
Windsor	.410	.462
Calgary	.418	.470
Vancouver	.579	.671
Toronto	.648	.729

Figure 3.2: Index of Qualitative Variation by Index of Diversity for 17 Cities

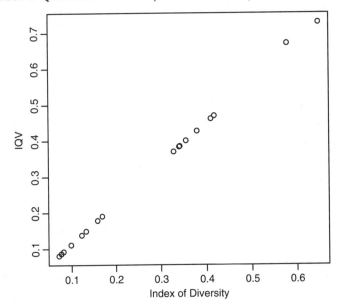

For the sample on the left, since the same number of people is found in each category, the IQV is 1.00. For the sample on the right, for the same reason, the IQV is also 1.00. For each, the message is that the dispersion is as great as possible. Yet there is clearly more variation in the sample on the right, because twice as many places of birth are represented.

The IQV is not responsive to the number of categories represented, only to the degree of dispersion across those categories. It gives us the levels of dispersion relative to the number of categories represented, but not a measure of sheer diversity.

An Information Theoretic (Entropy) Measure of Dispersion

A measure which is sensitive to the number of categories into which a set of cases is divided comes from information theory. To see its rationale, we shall re-examine country of birth, beginning with the four-country sample.

Canada	2
India	2
Nigeria	2
United States	2

We have two cases in each of four categories. Suppose we want to know in which category a given case is located. One bit of information (that is, one binary digit) will tell us whether it is in the upper pair of categories or the lower pair. An additional bit will tell us whether, within the pair, it lies on the top or the bottom, so two bits will tell us where it lies.

Suppose now that we have eight countries:

Canada	2	Nigeria	2
Britain	2	Mexico	2
Germany	2	Somalia	2
Hong Kong	2	United States	2

Here we need a further bit of information to distinguish the set of four categories on the left from the set on the right. Once we know that, we are down to four categories, and, as just seen, we need two bits of information to distinguish among four. Altogether, then, we will need three bits of information to eliminate our uncertainty. In general, if we double the number of categories, we need another bit.

Our uncertainty about where cases lie is a reflection of how widely they are dispersed. Our uncertainty is measured by the number of bits we need to wipe it out. Thus the number of bits needed to place the cases in their categories becomes a measure of dispersion.

Knowing where a case is located, we can express our knowledge with a binary code. For example, to locate a case in the first of eight categories, we can use the code 0 0 0, where the first "0" means that it is in the first set of four categories, the second "0" means it is in the first pair of cases within that set, and the third "0" means that it is in the first category within the pair.

Ordinarily, though, we can use codes that are more efficient. For example, if a great many cases are in a single category, we may assign it a one-digit code. If this category comes up often enough, we may be able to get by with fewer bits of information, on average.

Suppose we have eight people, born in different countries, as shown:

Canada   4

Britain   2

India   1

Nigeria   1

We can set up this code:

Canada   1

Britain   0   1

India   0   0   1

Nigeria   0   0   0

Although we are using up to three digits, on average this code will be more efficient than one with all two-digit codes. To see this, we just need to work out the mean number of digits required.

4 cases (from Canada) use 1 digit	**4(1) =**	**4**
2 cases (from Britain) use 2 digits	**2(2) =**	**4**
2 cases (from India and Nigeria) use 3	**2(3) =**	**6**
		14

$$\frac{14}{8} = 1.75$$

On average, we need 1.75 digits, rather than the 2 we needed when we treated all categories as being of the same size.

Fortunately, we do not have to work out a code for every variable to see how many digits we need. We can use Shannon's (1948) formula:

$$-\Sigma \, p_i[\log_2(p_i)] \, ,$$

where p_i is the probability of the ith category, and $\log_2$ refers to logarithms to the base of 2. (If you are part of the post-logarithmic generation, you may wish to look at the notes on logarithms in Appendix B.) In our example,

$$\text{p(category 1)} = \frac{4}{8} = .500, \ \log_2(.500) = -1$$

$$\text{p(category 2)} = \frac{2}{8} = .250, \ \log_2(.250) = -2$$

$$\text{p(category 3)} = \frac{1}{8} = .125, \ \log_2(.125) = -3$$

$$\text{p(category 4)} = \frac{1}{8} = .125, \ \log_2(.125) = -3$$

Substituting in $-\Sigma \ p_i \ [\log_2(p_i)]$, we obtain

$$- \ \Sigma \ .5(-1) + .25(-2) + .125(-3) + .125(-3) =$$

$$- \ \Sigma \ -.5 -.5 -.375 -.375 = -(-1.75) = 1.75$$

This is the number of digits we saw above that we needed.

Shannon's formula provides the number of bits of information we need, on average, to eliminate uncertainty about where cases are found. In the sense of information theory, this uncertainty is a measure of entropy (the degree of randomness in the data). That is why this statistic is referred to as a measure of entropy.

It is particularly helpful for multi-category variables with widely different proportions in the categories. For such data, we have no immediate sense of the extent of diversity. For example, let us look again at the occupational distributions of males and females, shown—this time in percentages—in Table 3.4.

Table 3.4: Occupational Categories of Canadian Males and Females, 2006, in Percentages

	Males	**Females**
Management	11.6	7.5
Business, finance, administration	9.7	27.1
Natural sciences	9.7	3.0
Health	2.1	9.6
Social sciences	5.2	12.1
Culture and recreation	2.5	3.5
Sales and service	19.3	29.1
Trades, transportation	26.7	2.2
Primary industries	5.7	1.8
Processing, manufacturing, utilities	7.5	4.1
	100.00	100.00

We will be hard pressed to say how much sheer diversity is present for either category without a statistic to express it. The entropy measures are 2.967 for males and 2.739 for females, suggesting, as did the Index of Diversity, that there is somewhat greater variation among males.

Unlike the IQV, the entropy measure does not give us the amount of dispersion in relation to a maximum. Rather, it is a measure of the absolute extent of diversity that is present. Because this is so, the two measures are complementary, and can be used together to highlight different aspects of the data.

Although the entropy measure is calculated quite differently from either the Index of Diversity or the IQV, all three attempt to measure the dispersion of nominal variables, and the entropy measure often turns out to be well correlated with the others. Figure 3.3 shows a strong association between the IQV and the entropy measure for the same Canadian cities examined above.

Figure 3.3: Entropy by IQV for 17 Cities

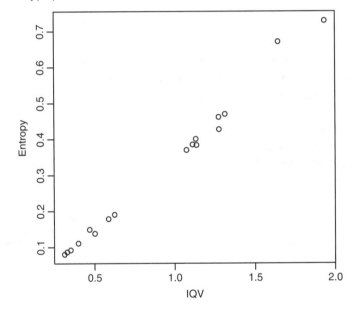

Although in this instance the measures are highly correlated, they have different conceptual meanings, and therefore can be used together.

Although it provides a measure of sheer diversity, the entropy measure does depend on how finely the data are broken down. In the occupational data presented above we have 10 categories. National data include 520 detailed occupations, and the entropy would surely be greater if we used that many categories. As it turns out, if we use all 520 categories, for males we need 7.841 and for females 7.136 bits, so the impression of greater dispersion for males holds up.

Quantile-Based Measures of Dispersion

Above the nominal level of measurement, we typically wish to use our ability to order the cases when we assess dispersion. We shall consider three measures which take advantage of this ability: the Interquartile Range (IQR), the Interdecile Range (IDR), and the Median Absolute Deviation (MAD).

With ordinal data, although we cannot specify precise distances between categories, we can say how much of the sample lies between particular values. We can tell our readers the upper and lower bounds within which the middle half or the middle 80% is located. With interval or ratio data, we can state the precise distance within which a fraction of the sample lies.

The most widely used measures for this purpose are the Interquartile and the Interdecile Ranges. Respectively, these give us the range between the upper and lower quartiles, and between the upper and lower deciles. Quartiles, unsurprisingly, are points that divide an ordered distribution into quarters. Deciles are points dividing an ordered set of cases into tenths. For example, suppose 20 students obtain the following grades:

51 59 61 63 64   66 66 67 68 69   71 72 74 75 77   79 80 81 83 85

Figure 3.4: Quartiles and Deciles

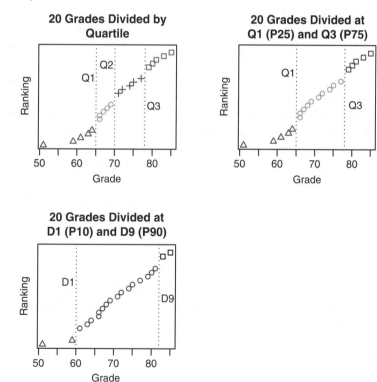

The upper left graph in Figure 3.4 splits the cases into four groups, divided at the quartiles. The first, or lower, quartile marks off the bottom quarter from the rest. It must lie between 64, the fifth grade, and 66, the sixth. When the dividing line falls between two cases, by convention, we place it halfway between them.[2] Following this strategy, we place it at 65. The second quartile, or median, cuts the top half from the bottom half. Here, it will lie halfway between 69 and 71, at 70. The third quartile divides the lower three-quarters from the upper quarter. Here it lies between 77 and 79, at 78.

The quartiles may be denoted Q_1, Q_2, and Q_3, or referred to in percentile notation as P_{25}, P_{50}, and P_{75}.

In defining the Interquartile Range (IQR) we compare the upper and lower quartiles, here at 78 and 65. Since we have ratio data, there is no problem with subtracting 65 from 78, and reporting the IQR as 13. This is just the range within which the middle half of the sample is to be found. If we were working with ordinal data, we might or might not choose to make the subtraction, but in either case we could report that the middle half of the class fell between 65 and 78. The IQR is illustrated above.

If we want to report the bounds of a larger fraction of our sample, the outer deciles can be used. The lower decile marks off the lower one-tenth from the rest, and the upper decile divides the upper tenth from those below. The deciles may be denoted D_1 and D_9, or put in percentile notation as P_{10} and P_{90}.

In the example above, including 20 students' grades, we would mark off the two lowest cases (at 53 and 59) from the rest, and also the two highest (at 83 and 85). (See Figure 3.4.) As with quartiles, when the dividing line lies between cases, we can, by convention, place it halfway between them. Here the lower decile lies between 59 and 61, so we place it at 60, and the upper decile between 81 and 83, so we place it at 82.

Since we have ratio data, we can subtract, and report the Interdecile Range (IDR) as $82 - 60 = 22$. If we had only ordinal data, and preferred not to subtract, we could still indicate that 80% of the cases were in the range from 60 to 82.

As we have seen, if we have interval or ratio data and we would like to report the range within which a fraction of our cases lies, this is fine. We may also choose the IQR or the IDR as a fallback when the standard deviation, the most common measure for interval or ratio data, is problematic.

An Alternative—The MAD

Another option sometimes suggested for ordinal data is the MAD (the Median Absolute Deviation). To get it, first we take deviations from the median and arrange them in order of their absolute values. Here we place the deviations from the median of 70 below the grades we have considered above.

51	59	61	63	64	66	66	67	68	69	71	72	74	75	77	79	80	81	83	85
−19	−11	−9	−7	−6	−4	−4	−3	−2	−1	1	2	4	5	7	9	10	11	13	15

In absolute value, the deviations are

19 11 9 7 6 4 4 3 2 1 1 2 4 5 7 9 10 11 13 15

Placed in ascending order, they are

1 1 2 2 3 4 4 4 5 6 7 7 9 9 10 11 11 13 15 19

Their median lies between the tenth value, 6, and the eleventh, 7, and is set at 6.5.

The MAD requires us to take absolute differences, that is, to subtract one value from another, then change the sign of the difference to positive if it is negative. If the data are interval or ratio, the subtraction is straightforward. If they are ordinal, the difference is an approximation. Notice, though, that the MAD, like the IQR and IDR, is highly resistant, that is, it is unaffected by extreme values. If we are concerned about the effects of extremes on the standard deviation, the most common measure for interval or ratio data, the MAD is thus a possible fallback.

The Standard Deviation

For interval or ratio variables, the most widely used measure of dispersion is the standard deviation.

It takes advantage of the clearly defined intervals between categories by working with the differences between individual observations and their mean. If the differences are small, it is entirely reasonable to say that the variable has a limited spread, and if they are large, a broad spread.

The difference between an individual observation and the mean, for the ith case, can be written

$$(x_i - \bar{x}) ,$$

where x_i denotes the value of x for the ith case, and $\bar{x}$ is the mean for all cases.

A second feature of the standard deviation is that it uses data from all of the cases. (Recall that the commonly used measures for ordinal data only use data from specific percentiles.) If we want to know how widely values of x are dispersed, we ordinarily want to look at them all. The obvious thing, at first glance, might appear to be to add up the differences between the x_i and their mean, but if we do positive and negative differences will cancel each other. (As shown in Chapter 2, they will cancel perfectly.) One way around this is to square the differences, obtaining

$$(x_i - \bar{x})^2$$

Having these, we can add them up, obtaining the *sum of squared deviations*, or the *sum of squares*, written

$$\Sigma\,(x_i - \overline{x})^2\,,$$

where the upper-case Greek sigma indicates that we are to add the squared deviations across cases.

Other things equal, we will get a larger sum of squared deviations for a large sample than for a smaller one. To get a measure of dispersion comparable across samples, we have to compensate. We divide by the number of cases in the sample. Denoting the sample size by N, we can write the result as

$$\Sigma\,(x_i - \overline{x})^2 / N$$

Since summing across cases, then dividing by N creates a mean, we can refer to this quantity as the *mean squared deviation*. More commonly it is called the *Variance*.[3]

There is one important difficulty with the Variance. When we square a quantity we also square its units. Think, for example, of an area 2 metres by 2 metres. The surface it covers will be $(2m)^2$, or 4 square metres. Here, squared units make sense, but suppose we want to examine the dispersion of incomes, and the Variance turns out to be, say, 5,000 square dollars. Squared units for income, and for almost all variables in social science, make no sense, so we need to desquare them. We do this by taking the square root of the Variance, which may be written as

$$\sqrt{\Sigma\,(x_i - \overline{x})^2 / N}\,, \text{ or } [\Sigma\,(x_i - \overline{x})^2 / N]^{.5}$$

This is the standard deviation, which is expressed in the original units of the variable in question. Because of its construction, it has four clear merits:

- it takes advantage of the well-defined intervals between categories of an interval or ratio variable;

- it uses information on all of the cases;

- it can be compared across samples; and

- it is expressed in meaningful units.

You will see the standard deviation denoted in various ways. The true standard deviation for a population is often referred to by the lower-case Greek sigma, σ. An estimate taken from a sample is often denoted by its counterpart from the Roman alphabet, the lowercase *s*. Sometimes it is referred to by its initials, as SD. You just need to recognize the various expressions employed.

Calculating a Standard Deviation

Today we let computers work out SDs. Nonetheless, it may be helpful to see how one once might have been calculated. An example which works out easily, allowing us to focus on the process rather than the calculation, is presented in Table 3.5.

To obtain $[\Sigma(x_i - \bar{x})^2 / N]^{.5}$ we set up a worksheet, shown in Table 3.5.

Table 3.5: A Worksheet to Obtain the Standard Deviation

x_i	$(x_i - 2)$	$(x_i - \bar{x}) = (x_i - \bar{x})^2$
0	−2	4
1	−1	1
2	0	0
3	1	1
4	2	4
$\Sigma x_i = 10$		$\Sigma(x_i - \bar{x})^2 = 10$
$\bar{x} = \Sigma x_i / N$		$Var(x) = \Sigma(X_i - \bar{x})^2 / N = 10 / 5 = 2$
$= 10 / 5 = 2$		$SD(x) = [Var(x)]^{.5} = 2^{.5} = 1.414$

Instability of the Standard Deviation

Since we square deviations from the mean to get the SD, outlying values can be highly influential, and the SD can readily become unstable. Consider this set of grades, graphed in Figure 3.5.

35 63 63 69   71 74 76 77   79 80 81 81   82 84 86 88

Figure 3.5: Histogram of Grades

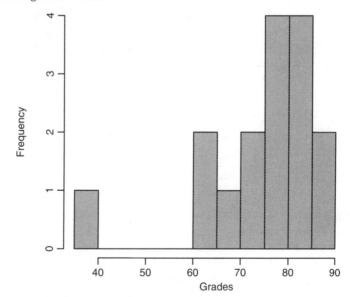

One student, who did not complete the work for the course, lies far below the others. How much difference does the outlier make?

With it, the SD = 12.81

Without it, the SD = 7.65

In cases like this, it is prudent to consider using the IQR, here 11.5, the IDR, here 22, or the MAD, here 7.

The Standard Deviation with Ordinal Data

One reason to use the SD with ordinal data is that measures often recommended, the IQR, IDR, and MAD, may show major changes when the shifts in the data are modest, or no changes when the shifts in the data are major.

Consider the graphs in Figures 3.6. and 3.7. Each is based on 100 cases. In the first graph in Figure 3.6, there are 9 in each of the outer categories. In the second, there are 11. Note that the IDR has tripled, going from 1 to 3. The MAD, on the other hand, shows no change at all. The SD shows a modest rise, from .78 to .84. In the third graph, there are 48 cases in each of the outer categories, but the IDR has remained stuck at 3, while the MAD has tripled. The SD has risen to 1.49.

Figure 3.6: Shifts in the SD, IQR, and MAD

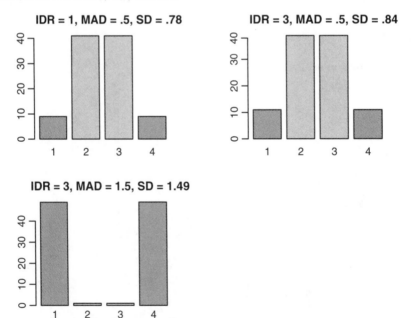

The graphs in Figure 3.7 bring in the IQR. In the first, the central categories are slightly larger than the extremes, and in the second the situation is reversed. There is little change in the data, but the IQR and the MAD have both tripled, while the SD has edged up from 1.11 to 1.14.

Figure 3.7: Further Shifts in the SD, IDR, and MAD

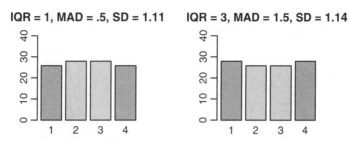

The SD responds much more smoothly to changes in the data than the IQR, the IDR, and the MAD. Many use the SD with ordinal data for that reason, as long as they can accept that the distances between categories are not too seriously uneven.

Standardizing Variables

Beyond its use as a measure of dispersion, the SD can be used to put variables into "standardized" form. To do this, we take

$$z_i = (x_i - \bar{x}) / SD(x)$$

The variable name has been changed from x to z, since standardized variables are referred to as z-scores, and denoted accordingly.

The resulting distribution is tidy, because

a) its mean is zero, and

b) its SD is one.

Why these results hold is shown in Appendix A2.

These simplifications allow us to compare the shape of two distributions without being distracted by differing means and SDs. Z-scores also let us see how far a given case lies from the mean in a simple way, by expressing the distance in standard deviations.

The Standard Deviation and the Tails of a Distribution

If we know the SD, we can say what proportion of cases will lie in the tails of a distribution, but the proportion depends on the distribution. If it is normal, 68% of cases will lie within one SD of the mean, and 95% within 1.96 SDs. (Often we, somewhat informally, use a two-SD approximation.) We shall see later that this fact is used heavily in statistical inference. These proportions, with some others, are illustrated in Figure 3.8.

Statisticians often speak of an "empirical rule," that many distributions encountered in practice, if not normal, still have roughly the same proportions within a given number of SDs from the mean. More specifically, for many distributions about 95% lie within two SDs of the mean and 99.7% within three. For example, the distribution of monthly household incomes in Figure 3.9 is skewed to the right, so there are no cases more than two SDs to the left of the mean, but 96.4% still lie within two SDs of the mean.

Figure 3.8: Regions Containing 68%, 95%, 99%, and 99.9% of a Normal Distribution

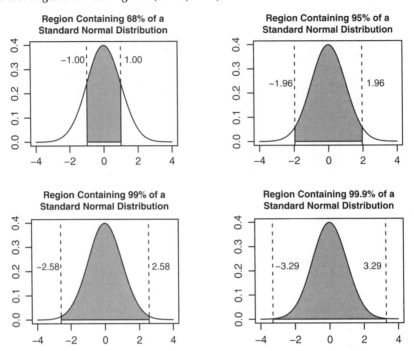

Figure 3.9: Household Income, 96.4% of Cases within Two SDs of the Mean

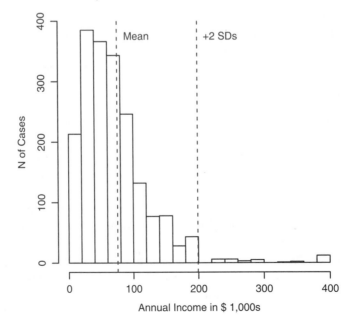

More technically, if a distribution is strictly continuous and unimodal, with a definable standard deviation, no more than about 11.1% can lie further than two SDs from the mean, and conversely 88.9% will lie within that range (Vysochanskii and Petunin, 1980).[4]

The category of continuous and unimodal includes a great many distributions, as illustrated in Figure 3.10. The percentage within two SDs of the mean is given for each graph. For each the percentage is at least 90, and apart from the first, which is quite heavy-tailed, the percentage is at least 95.

Figure 3.10: Some Unimodal Continuous Distributions

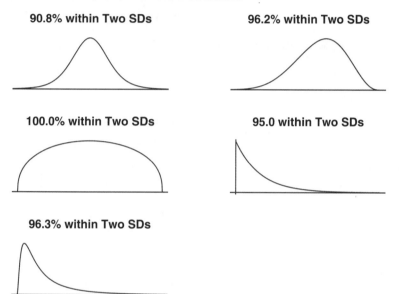

90.8% within Two SDs

96.2% within Two SDs

100.0% within Two SDs

95.0 within Two SDs

96.3% within Two SDs

The Special Case of the Dichotomy

The formula for the SD simplifies for dichotomies. If we score the two categories 0 and 1, and define p to be the proportion of 1s, or the probability of getting a 1 on a random draw, then the mean equals p. The standard formula is

$$\overline{x} = \sum x_i / N$$

Here, since x takes on only the values 0 and 1, we can write

$$\overline{x} = [(\# \text{ of } 0s)*0 + (\# \text{ of } 1s)*1] / N$$

Since (# of 0s)*0 = 0, and (# of 1s)*1 = (# of 1s), we can simplify to

$$\overline{x} = (\# \text{ of } 1s) / N$$

The number of 1s over the number of cases gives us the probability of a 1, so $\overline{x} = p$. Since cases score either 0 or 1, the probability of a 0 is just 1 − p. This quantity, 1 − p, is often called q. Straightforwardly, p + q = 1.

Since the SD is the root of the Variance, we can easily obtain the SD if we have the Variance. Its standard formula is

$$\Sigma\, (x_i - \bar{x})^2\, /\, N$$

Inserting p for the mean of x, we get

$$\Sigma\, (x_i - p)^2\, /\, N \qquad\qquad [1]$$

The difference between x_i and p must be either $(0 - p)$ or $(1 - p)$, so we can break the formula in two.

$$\Sigma\, (x_i - p)^2\, /\, N$$

equals the sum, across the 0s, of $(0 - p)^2\, /\, N$,

plus the sum, across the 1s, of $(1 - p)^2\, /\, N$

We can write this as

$$\frac{(\text{\# of 0s})^*(0 - p)^2}{N} + \frac{(\text{\# of 1s})^*(1 - p)^2}{N}$$

Recall that (# of 1s) / N = p. Also (# of 0s) / N = (1 − p), or q. Substituting, we obtain

$$q(0 - p)^2 + p(q)^2$$

Expanding $(0 - p)^2$ gives us $[0^2 + 0(-p) - p(0) + p^2] = p^2$.

Substituting yields

$$qp^2 + pq^2\ ,\ \text{ or }\ p^2q + pq^2$$

Factoring leaves

$$pq(p + q)$$

But $(p + q) = 1$, so we have

$$pq(1) = pq$$

Given this result for the Variance, the SD is

$$\sqrt{pq}\ \text{ or }\ [pq]^{.5}$$

From this formula, if we have the mean, p, we can readily obtain the SD. In some journals in some disciplines, people only report the mean of a dichotomy, because the SD is a simple function of p. In social science, we typically report both p and its SD, but we should be aware that, as p, the proportion of 1s, departs from .5 the SD declines regularly, as shown in Figure 3.11.

$$\Sigma\, (x_i - \bar{x})^2\, /\, (N)$$

Figure 3.11: Standard Deviation of a Proportion

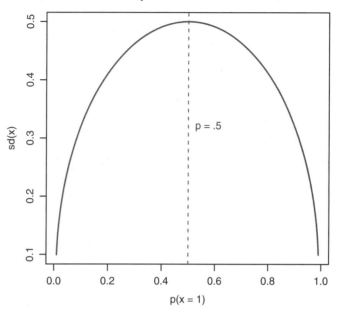

Summary

In this chapter we have surveyed measures of dispersion designed for nominal, ordinal, and interval or ratio data. Two of the measures for nominal variables are based on whether cases are matches, that is, on whether they are in the same category. The Index of Diversity provides the probability that two cases chosen at random will be in different categories, and the Index of Qualitative Variation (IQV) gives us the number of unmatched pairs as a proportion of all possible unmatched pairs. The third measure appropriate for nominal data is based on entropy and tells us how much information we need to eliminate uncertainty about category membership.

When dealing with ordinal data, we typically make use of quantiles, as in the Interquartile Range (IQR) and the Interdecile Range (IDR). We can also use the Median Absolute Difference from the median (the MAD).

With interval or ratio data (or ordinal data we prefer to treat as interval) the standard deviation is typically employed. We have seen that it is based on our ability to say how far cases lie from the mean. Beyond its use as a measure of dispersion, the standard deviation is used in creating z-scores. As well, for a given distribution, we can say how many cases lie how many SDs from the mean. This fact will be heavily used later when we discuss statistical inference.

Review Questions on Measures of Dispersion

1. When might we prefer to use an entropy measure of dispersion rather than an IQV? Rather than a standard deviation?

2. In the formula $[N^2 - \sum n_j^2] / [N^2 - N^2/k]$, what do N, n_j, and k stand for? What does this formula give us?

3. Suppose we see the formula $-\sum p_i[\log_2(p_i)]$. What do p_i and $\log_2$ stand for? If we calculate this quantity, what will it tell us?

4. Suppose someone said to you that there is no measure of dispersion for nominal (ordinal) variables, because dispersion is meaningless when we cannot tell how far apart categories are. What might you say in reply?

5. In what way are the IQV and the entropy measure complementary?

6. What measures of dispersion are commonly suggested for ordinal variables? Why, for truly ordinal variables, may it be safer just to report key percentiles?

7. Suppose that the IDR for the final grades in a course in social statistics was found to be 21. What would this tell us about the distribution of grades? If the MAD was 10, what would this tell us?

8. What measure of dispersion is typically suggested for ratio variables, and when is it liable to be misleading?

9. What is the formula for a standard deviation? Why is there a square root sign in the formula?

10. Explain the meaning of the symbols in the formula for the SD.

11. Briefly state the advantages of the standard deviation.

12. When might we prefer to use an IQR rather than a standard deviation?

13. Why do many researchers prefer the SD to the IQR or IDR (or MAD) when they have ordinal data?

14. What are the mean and SD of a z-score? What are two ways z-scores can be helpful?

15. If a variable is normally distributed, what percentage of observations will lie within approximately two SDs of the mean? Within one SD?

16. What is the "empirical rule"?

17. If a variable is strictly continuous and unimodal, what percentage of observations will lie within two SDs of the mean?

Notes

1. The indices were calculated from the raw numbers rather than the percentages shown here.

2. Statistical packages often use more elaborate methods, based on the idea that, with more precise measurement, observations would be distributed around the specific values we have. Depending on the assumed distributions, different quartiles are obtained. The same holds for deciles.

3. For the variance and standard deviation of a sample, we ordinarily use $(N - 1)$ rather than N. The difference is of no great importance except for small samples, and will be neglected here.

4. The only distribution well known in the social sciences for which more than 11.1% of observations lie more than two SDs from the mean is the t distribution, with fewer than five degrees of freedom. With four, 11.6% are more than two SDs from the mean; with three, 13.9%; and with two, 18.4%. With one, the SD cannot be defined.

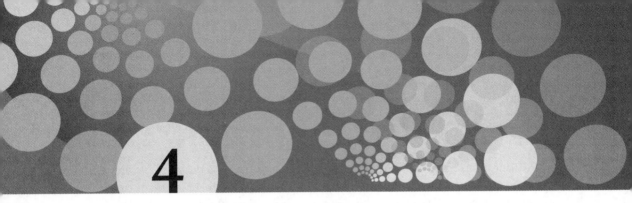

4

Describing the Shape of a Distribution

Learning Objectives

In this chapter, you will learn

- ways to describe a distribution, in terms of the number of peaks in its graph, the direction in which its tails point, and the level of its central peak;

- a series of distributions that have special names, based on the shape of their graphs; and

- the limits of the formula used most often to assess whether the graph is skewed in one direction or the other, and of the formula designed to tell us about the height of a central peak.

Although measures of central tendency and dispersion are very helpful in describing a distribution, they are clearly not sufficient. To illustrate, Figure 4.1 shows scores on a widely used measure of depression, the Center for Epidemiological Studies (CES) scale, which is based on the extent of symptoms. Most people have few and infrequent symptoms, and the numbers fall off as scores rise, but a modest proportion have scores in the 60s and 70s, which suggest severe problems.

Given only the means and standard deviations we could not have known how many had scores at any given level, because very differently shaped distributions can have the same means and SDs. Figure 4.2 illustrates.

Figure 4.1: CES Depression Scores

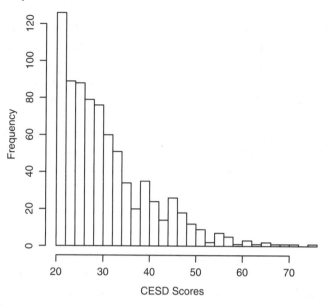

Figure 4.2: Same Mean and SD

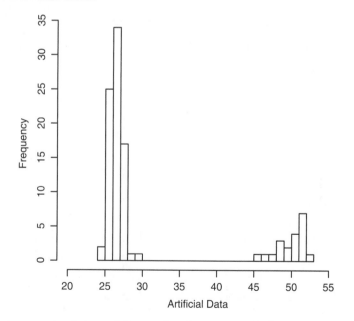

If these data were real, it would seem that there were two distinct groups of people, a larger one with modest symptoms and a smaller one with significant issues. However far from reality this graph may be, it is based on the same means and SDs as the real data.

We must be aware of the shape of a distribution as well as its central tendency and dispersion.

This section will make less of the distinctions among levels of measurement than the previous ones. The reason is that the shape of the distribution for a nominal variable is arbitrary. If there is no compelling reason to code its categories in any particular order, there is no reason why they should appear on a graph in any particular order. Most of the terms, and the measures we use to describe the shape of a distribution, apply only to ordered variables.

We shall look, in turn, at the number of peaks appearing in a graph, at the length of tails on the left and right, and at the height of central peaks. In Chapter 5, we shall look at graphic presentation of distributions, and at some ways to summarize them succinctly.

Modes

As just pointed out, since the graphs for nominal variables are arbitrary, we usually say little about their shape. The exception lies in terms for the number of modes observed. The term "mode" here bears a different sense than it does when we are describing central tendency. There the mode is the most frequent category. In describing distributions, a mode is a pronounced peak in a graph. If it displays one mode, as in Figure 4.3, which shows year of birth for voting-age Canadians, we describe it as unimodal.

Figure 4.3: Year of Birth for Voting-Age Canadians, 2008

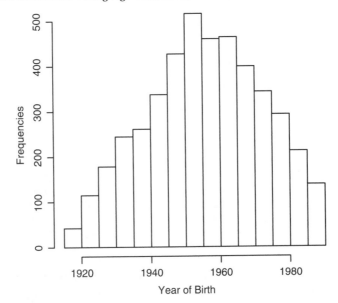

We describe it as bimodal if there are two clear peaks, as in the example in Figure 4.4. This graph represents a scale intended to pick up whether people tended to talk to themselves when remembering events. Many never did and many did for every event they were asked about. (This graph came from exploratory work by Arnold, looking at whether how people remember things affects mathematics phobia.)

Figure 4.4: Histogram of Self-Talk

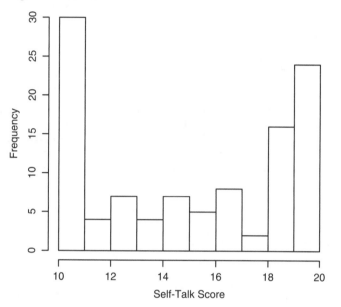

In the more extreme situation, in which there are three or more pronounced peaks, we refer to the distribution as multimodal.

Skewness and Kurtosis

The two other features of a distribution on which we most often remark, skewness and kurtosis, do not apply to nominal variables, because they require the categories to be meaningfully ordered. Skewness has to do with the balance between the tails of a distribution. If the right tail is pronounced compared to the left, we speak of rightward (or positive) skewness. This is illustrated by scores on the Center for Epidemiological Studies Depression scale, shown in Figure 4.5.

If the left tail outweighs the right, we speak of leftward, or negative, skewness, as in the scores on a scale of mathematics phobia, applied to a class in sociology and shown in Figure 4.6.

Figure 4.5: Histogram for CES Depression Scores

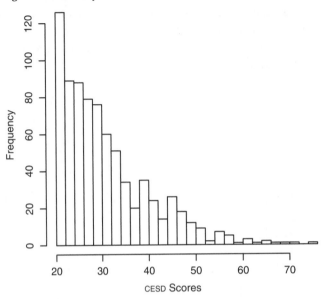

Figure 4.6: Histogram of Math Phobia

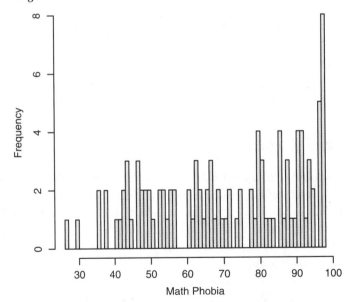

If the left and right tails are in balance, we speak of the distribution as symmetric. This description fits the standard normal curve, shown in Figure 4.7.

Figure 4.7: Standard Normal Curve

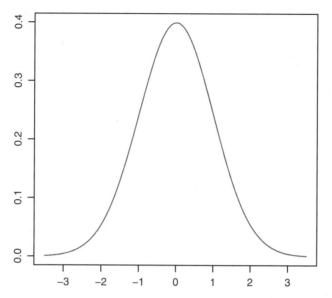

Kurtosis, like skewness, is assessed only for ordered distributions. By definition, the standard normal curve is "mesokurtic," meaning that its central peak is of medium height. A distribution with a high central peak is said to be "leptokurtic." An example is the income distribution shown in Figure 4.8, with a portion of the normal curve superimposed.

Figure 4.8: Annual Household Income, Compared to a Normal Distribution

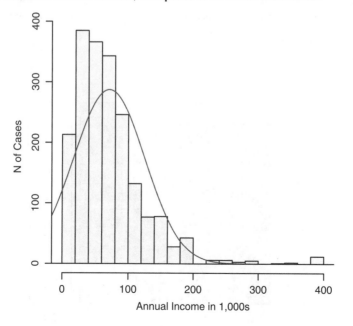

Note that to speak of a central peak, we must have cases on each side of it, so unless there are two tails the concept of kurtosis is inapplicable. Note also that the concept of kurtosis does not apply well when there is more than one peak in a distribution—i.e., when we have bi- or multimodal data.

A distribution with a low central peak, like the one in Figure 4.9, is "platykurtic."

Figure 4.9: Responses to "I Seek Clarity"

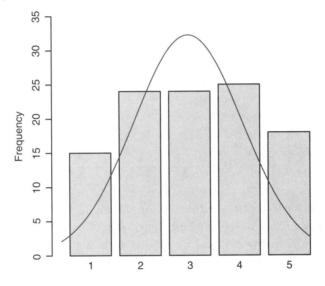

Where there is no peaking, we have a uniform, or rectangular, distribution. We do not often see precisely uniform distributions in practice, but some come close, as in Figure 4.10.

Figure 4.10: Responses to "I Am a Private Person"

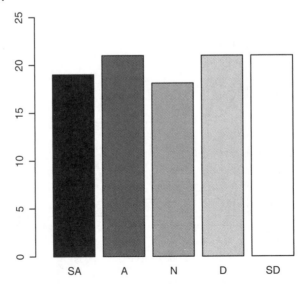

Some special shapes are approximated often enough to have their own names. For example, distributions may be J-shaped, as illustrated in Figure 4.11,

Figure 4.11: Ratings of Marital Quality

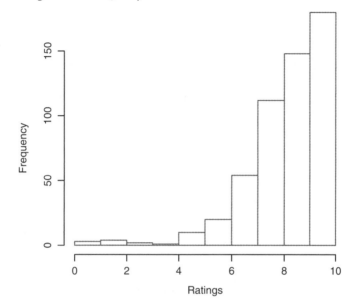

or reverse J-shaped, as in Figure 4.12,

Figure 4.12: Number of Evening Outings to Restaurants and Bars in the Previous Month

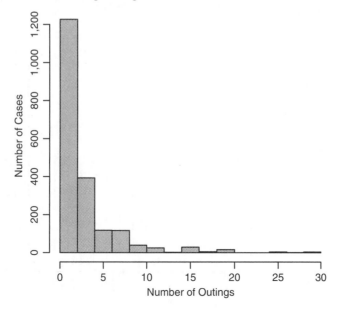

or L-shaped, as in Figure 4.13, showing how much of the work in an introductory course one year's students did not attempt.

Figure 4.13: Value of Exams/Papers Not Attempted

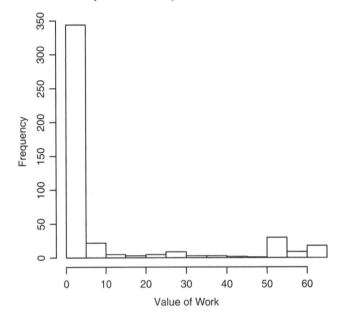

The vast majority omitted work worth less than 5% of the final mark, but others left out varying amounts.

Other distributions, like the one in Figure 4.14, may be described as U-shaped.

Figure 4.14: Histogram of Cases from a U-shaped Distribution

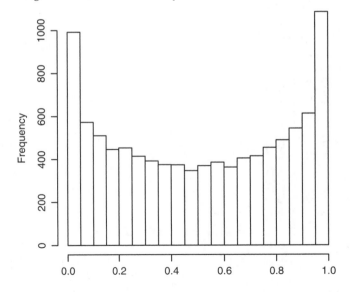

Formulae for Skewness and Kurtosis

We see numeric measures for skewness and kurtosis less often than those for central tendency and dispersion. One reason is that we focus on skewness and kurtosis less often. Another is that the shape of a distribution is often best shown graphically. A third is that measures of skewness and kurtosis are not always satisfactory. Nonetheless, we ought to see what they attempt to show us, and then how they can go wrong.

The most common formulae are still those defined by Pearson. The most widely used measure for skewness can be written as

$$Sk = \frac{\Sigma \, (x_i - \bar{x})^3 / \, N}{Var(x)^{1.5}}$$

For kurtosis, he suggested another which can be written as

$$Ku = \frac{\Sigma \, (x_i - \bar{x})^4 / \, N}{Var(x)^2}$$

These measures can be very unstable in the presence of extreme scores. In the formula for skewness, we use $\Sigma (x_i - \bar{x})^3$, and for kurtosis $\Sigma (x_i - \bar{x})^4$, so outliers can have great influence. We shall look at examples in a moment.

In the measure for skewness, we cube the differences between individual observations and the mean. If $x_i > \bar{x}$, the cubed term $(x_i > \bar{x})^3$ will be positive. If $x_i < \bar{x}$, it will be negative. If the distribution is symmetric, the positive and negative terms for individual cases will cancel, leaving a sum of zero, which informs us of the symmetry. If the distribution has a long right tail, then the cubed terms for the cases with $x_i > \bar{x}$ will outweigh those for cases where $x_i < \bar{x}$, and the sum for all cases will be positive. If there is a long left tail, the sum will be negative. Thus Sk attempts to identify the direction of skewness.

Unfortunately, this measure is not expressed in meaningful units,[1] so it can be used only as an indicator of the direction of skewness, with higher values implying greater skewness. Even without meaningful units, we can get a sense of what Sk tells us from examples. Distributions with rising scores, beginning at 1.09, are shown in Figure 4.15.

Because of the great influence of outliers on Sk, it is good to examine a graph before using it. To see why, consider the two instances in Figure 4.16. In the first, because of an outlier on the left, Sk = 0, suggesting the absence of skewness, for what is otherwise a plainly right-skewed distribution. In the second graph, where there is no obvious outlier, a light left tail, which happens to go farther from the mean of 0, balances the effect of a heavier but shorter right tail. Again Sk suggests symmetry where that is not what we observe. With distributions like these we are better presenting a graph than using Sk.

As is the case for skewness, the formula for kurtosis is not expressed in meaningful units. For kurtosis, it is common to make a comparison to the normal distribution. To do this, 3, the score for a normal distribution, is subtracted from the basic formula. The normal thus finishes at 0. By definition, it is mesokurtic, so figures close to 0 imply a mesokurtic distribution, figures well above a leptokurtic distribution, and figures clearly below a platykurtic distribution.

Figure 4.15: Distributions with Varying Skewness

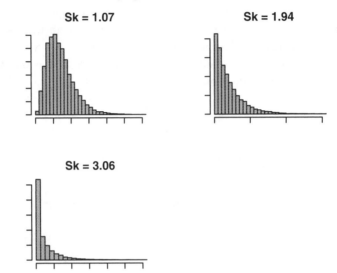

Figure 4.16: Two Distributions with Sk = 0

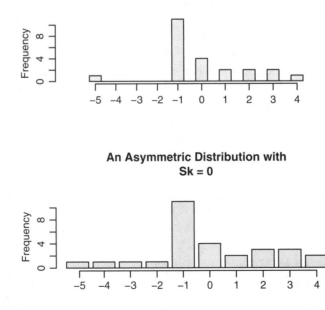

Ku responds much more dramatically to high peaks than to low ones. It can rise to infinity, but a perfectly rectangular distribution comes in at only −1.2. Although the concept of kurtosis refers to a central peak, the formula yields a figure for a distribution with a central trough. In this case some speak of "negative kurtosis." In the extreme of this phenomenon, when we have equal numbers at the far ends of the distribution and nothing in between, Ku declines only to −2.75. For illustrations of how differently Ku treats high and low (or negative) peaks, see Figure 4.17.

Figure 4.17: Distributions with Varying Kurtosis

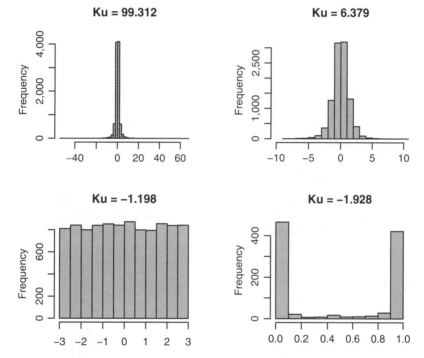

As with Sk, it is good to examine a graph before using Ku, because of the influence of outliers. Figure 4.18 shows an instance in which a plainly platykurtic distribution, with the addition of two outliers, moves to a Ku of 3.01, almost exactly that of the normal.

One potential solution to the influence of outliers is to use a measure based on percentiles. Several have been suggested, all resistant to outliers, but these are not commonly seen in social science today, so they will be passed by.

In suggesting one percentile-based measure, Moors (1986) also raised a conceptual issue with Ku. Moors showed that a high score can result from a high central peak, from heavy tails, or from a combination. The two distributions in Figure 4.19 illustrate. The one on the left has longer tails, and the one on the right a sharper central peak, but both have Ku = 3. (So we can focus on Ku, the two have identical means, SDs, and values of Sk.)

Figure 4.18: An Illustration of the Effects of Outliers on Ku

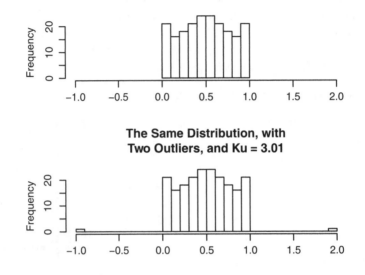

A Platykurtic Distribution, with Ku = −1.094

**The Same Distribution, with
Two Outliers, and Ku = 3.01**

Figure 4.19: Two Distributions with Ku = 3

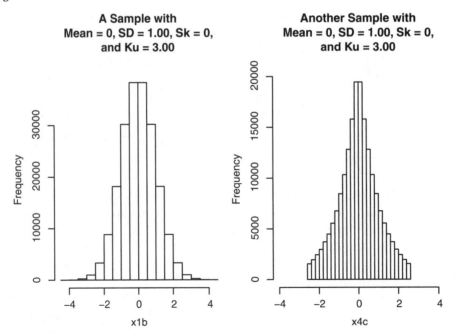

**A Sample with
Mean = 0, SD = 1.00, Sk = 0,
and Ku = 3.00**

**Another Sample with
Mean = 0, SD = 1.00, Sk = 0,
and Ku = 3.00**

Because we cannot tell from Ku whether the central peak is high or the tails are heavy, or both, it is advisable to examine a graph when an unusual value of Ku appears.

Summary

This chapter has reviewed ways to describe the shape of a distribution: the number of modes, which is applicable whatever the level of measurement, then skewness and kurtosis, which apply only to ordered distributions. We have also looked at a series of shapes which are common enough to have their own names: J-shaped, reverse J-shaped, and so on.

We have noted that the standard formulae for skewness and kurtosis are vulnerable to outliers, so that when we are concerned about this it is often helpful to examine a graph. If necessary, we can present one to show the shape of the distribution. Many graphic methods will be reviewed in the following chapter.

Review Questions on Shapes of Distributions

1. Draw a leptokurtic rightward skewed distribution. A platykurtic symmetric distribution. A bimodal distribution.

2. What does kurtosis refer to? Skewness?

3. What has level of measurement to do with kurtosis and skewness?

4. For kurtosis to make sense, what does a distribution have to have (besides ordered categories)?

5. In statistical terminology, what is the difference between a reverse J-shaped distribution and a uniform distribution? Between a U-shaped distribution and a normal distribution?

6. Suppose we read that the skewness of one distribution is −.50, and that of another is 1.00. Why might these be misleading? If the measures are meaningful for the data, what do these figures tell us?

7. Suppose we read that the kurtosis of a distribution is .50, and that of another is −2.50. Assuming the data are well behaved, what do these figures tell us?

8. What conceptual problem did Moors find with a standard measure of kurtosis?

Note

1. In the denominator, we have squared units, usually hard to interpret with social science data, and here they are raised to the power of 1.5.

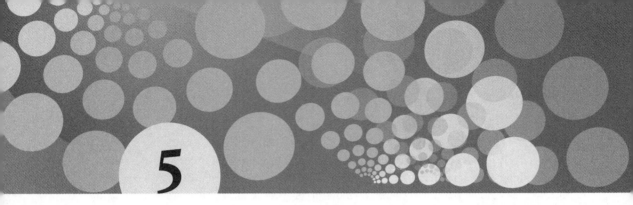

Summarizing a Distribution

Learning Objectives

In this chapter, you will learn

- a simple way to summarize a distribution with five numbers;
- its graphic counterpart, the boxplot;
- other graphic methods used to display a distribution, including pie charts, bar charts and histograms, and dot charts;
- extended uses of the bar chart; and
- a plot to display a cumulative distribution.

The Five-Number Summary

Besides referring to central tendency, dispersion, and shape, we can summarize an ordered distribution with a five-number summary, or a boxplot. The five-number summary uses

- the minimum,
- the first quartile,
- the median,
- the third quartile, and
- the maximum.

In doing so, it reduces the complexity of the data to something we can quickly grasp. Consider the distribution of years of birth graphed in Figure 5.1.

Figure 5.1: Year of Birth for Voting-Age Canadians, 2008

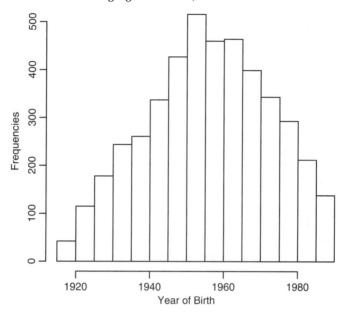

We have data on 4,429 cases, broken down by single years of age. We could not keep this much information in our minds if we wanted to, but we can readily grasp the five-number summary:

- minimum 1915
- Q1 1945
- median 1957
- Q3 1969
- maximum 1990

Typically, though, we would not see it spelled out this way. We would just see the numbers 1915, 1945, 1957, 1969, 1990, or perhaps we would see a boxplot.

Graphing a Distribution

Boxplots

The boxplot corresponds to the five-number summary in that it displays the minimum, the three quartiles, and the maximum. In the boxplot, a central box covers the middle half of the distribution, with its lower end at the first quartile and its upper end at the third.

The median is indicated by a line across the box. For an initial illustration, see Figure 5.2.

Figure 5.2: Annual Household Income, Canada, 2007

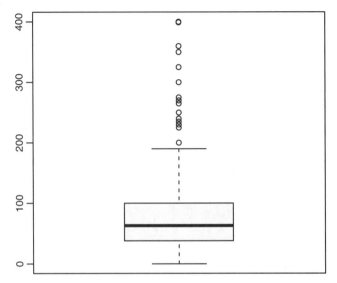

Outlying cases are indicated by circles or dots. By convention, a case is treated as an outlier if it is more than 1.5 IQRs from the nearest quartile. For normal or similar distributions, about 1% of cases will be identified in this way, and there will likely be about as many outliers at the top as at the bottom. For a strongly skewed distribution, there will be more on one end, as illustrated in Figure 5.3.

Figure 5.3: Histograms and Boxplots for a Normal and a Skewed Distribution

**Histogram of Cases for a
Normal Distribution**

**Boxplot of Cases for a
Normal Distribution**

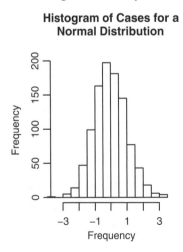

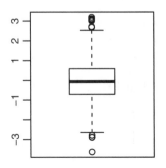

Figure 5.3 continued

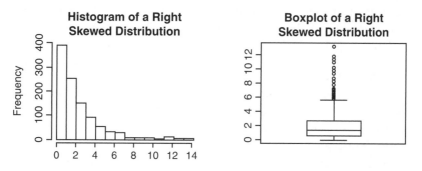

A long-tailed distribution will have many outliers on each side of the central box, as illustrated in Figure 5.4.

Figure 5.4: Histogram and Boxplot for a Long-Tailed Distribution

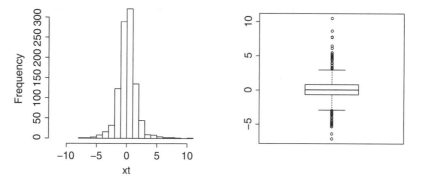

The boxplot's ways of displaying skewness are most readily seen when one is placed on its side for comparison with a histogram, as in Figure 5.5.

Figure 5.5: Histogram and Boxplot for the Same Data, Annual Household Income, Canada, 2007

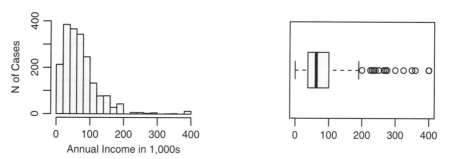

Notice that incomes over $193,000 are represented as outliers, that the whisker on the right side of the box is longer than that on the left, and that within the box the median is a bit to the left. In these three ways the boxplot points to the right skew of the distribution.

A Point of Interpretation

If more than a quarter of the cases take on the same value, the quartiles may overlap, and two lines in the box may occupy the same location. This result is often shown by a thicker than usual line. In Figure 5.6, the single line at the bottom shows that the first and second quartiles for females are the same.

Clearly, the five-number summary and boxplot reduce a complex array of numbers to something much easier to grasp. The boxplot, for the same reason, makes it straightforward to compare two distributions. Below we see that at least half the females attended no sporting events (because the median lies at zero), and their third quartile is about the same as the male median. In Figure 5.7 we compare levels of math phobia for two groups of students, those who did not report a sharp drop in liking for math before Grade 11 and those who did. Plainly those who did, on the right, are higher at every quartile, as well as at the very bottom of the range. The median for those who did is above the third quartile for those who did not.

Figure 5.6: Number of Evenings at Sporting Events, by Sex

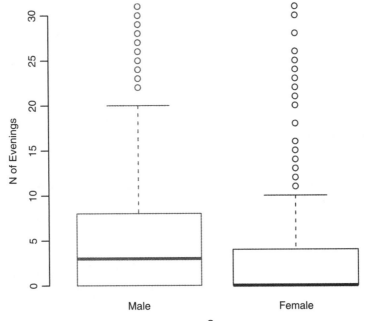

Figure 5.7: Math Phobia by Whether R Reports a Previous Drop in Liking for Math

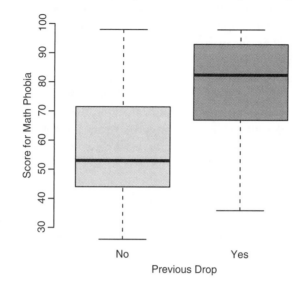

Pie Charts, Bar Charts, and Histograms

Rather than providing information about quartiles and outliers, we often wish to show how many cases are found in specific categories. How the graph appears can then depend greatly on how finely the categories are broken down. In Figure 5.8 we see two graphs based on a scale intended to measure dislike of working with computers, or "computer phobia." Even though based on the same cases, the second graph is multimodal and much less regular in its outline.

Figure 5.8: Histograms of Computer Phobia with Different Category Widths

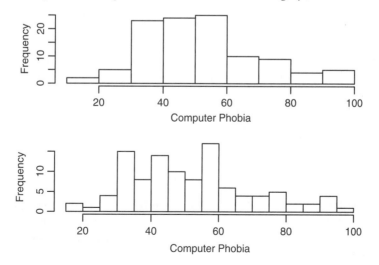

In choosing the number of categories (or "bins") to use, and where to draw the boundaries between bins, we try to present the essential features of the data without distracting readers with unnecessary detail. In Figure 5.9 we can consider three versions of year of birth, with intervals of ten years, five, and two.

If we wish to show how many people were born in specific decades, the first graph will be helpful. The second, using five-year categories, provides a clearer view of the shape of the distribution. The third, with two-year categories, provides a more detailed view but displays many noticeable irregularities of the kind that could result from random fluctuations. A graph with single-year categories is not presented. Readers who want year-to-year detail might be better viewing a table, where precise figures are more readily apparent than in a graph.

Figure 5.9: Year of Birth with Three Category Widths

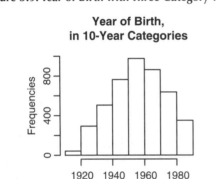

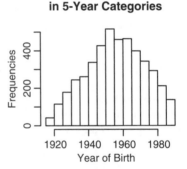

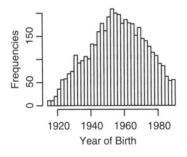

Besides deciding on the number of categories to use, we must often decide among pie charts, vertical histograms (or bar charts), and horizontal histograms (or bar charts). An illustration will help in considering their relative merits. We shall work with the number of people at each of the sites in a study in which the author was involved. First, in Figure 5.10 we see a basic pie chart.

Pie charts are widely familiar, but viewers may not find it easy to estimate the proportions in each category, or to rank them. Some research suggests that the eye is better at comparing the length of lines than at comparing angles or areas (Cleveland and McGill, 1985, 1987). For example, in Figure 5.10 can you readily tell which category is second largest?

We can, of course, place percentages beside the slices, or within the slices, as shown in Figure 5.11. While the relevant data are present, the viewer must look for the percentages within the graph. It might be as easy to take them from a table.

Figure 5.10: Distribution of Cases across Sites

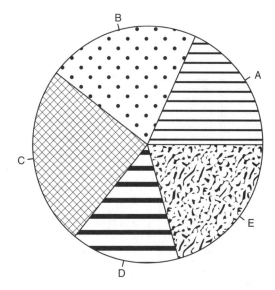

Figure 5.11: Distribution of Cases across Sites, Showing Percentages

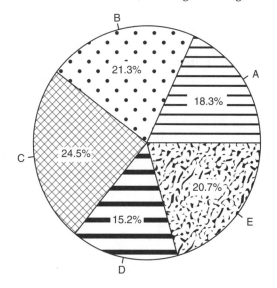

While it is easy enough to see the relative size of groups if there are few of them, and they differ greatly in size, pie charts are not the graph of choice for clarifying the relative sizes of groups differing by relatively little. Comparison is usually easier with a bar graph.

Since the bars are proportional in area to the number of cases they represent, they give a sense of the distribution of cases. Consider Figure 5.12.

Figure 5.12: A Vertical Bar Chart Showing Distribution of Cases across Sites

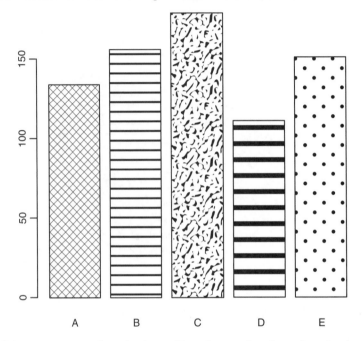

Some literature suggests that a horizontal bar chart makes the task easier than the more familiar vertical graph, because the eye can judge the length of horizontal lines more readily than vertical. Does Figure 5.13 make the comparison easier?

The Dot Chart

If we only want to show the relative number of cases in a group, we can do so with less ink through a dot chart, as in Figure 5.14. Fine lines allow us to link category names to the dots, without taking up nearly as much space as would be used for a barplot. On the other hand, if a graph must be visible from a distance, or must have visual impact, the dot chart may not be the preferred choice.

Figure 5.13: A Horizontal Bar Chart for Distribution of Cases across Sites

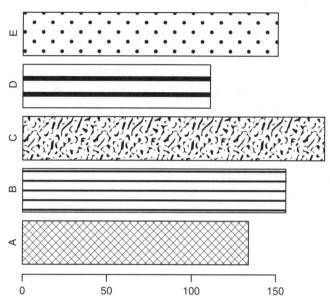

Figure 5.14: A Dot Chart Showing Distribution of Cases across Sites

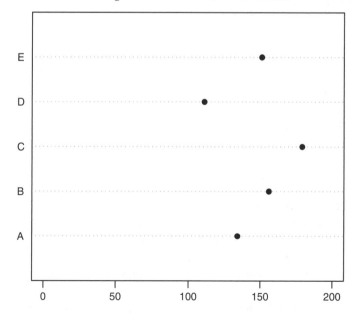

Smoothed Histograms

If it is necessary only to convey the general shape of a distribution, we can choose a smoothed curve, perhaps an Average Shifted Histogram (ASH), as shown in Figure 5.15. To get an ASH we use narrow bins and shift their boundaries gradually. At each step we record the height of the bar in each bin, and then average the heights from each step. In the upper left panel below is an ASH for the year of birth data we have examined above. For these data, the shape of the distribution is clear enough without smoothing, but in other situations smoothing can clarify an irregular graph, or resolve differences among graphs which look quite different.

Figure 5.15: An ASH for Year of Birth

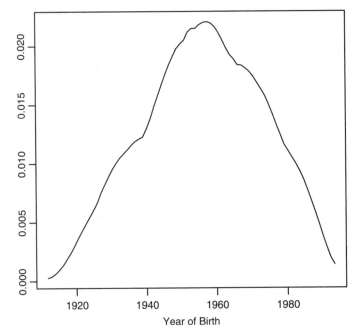

Smoothing may be particularly helpful when we wish to compare the shapes of several distributions. In Figure 5.16 we have scores on the Center for Epidemiological Studies Depression scale for the same people at three different times.

From the histograms we can get a sense of rough similarity over time, with specific variations, but we need to compare three graphs. The single graph in Figure 5.17 allows quick comparison, and likewise leads to an impression of broad similarity. It was produced through kernel smoothing, in which data points contribute to the height of the smoothed line not just at their own x-values, but at nearby values of x as well. Their contribution declines to zero as the distance from their own values of x increases. This approach produces graphs which can be overlain, and which often look much like those created by ASH routines.

Figure 5.16: Depression Scores on Three Occasions

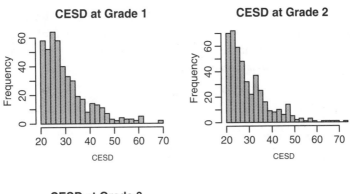

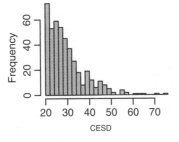

Figure 5.17: Depression Scores at Times 1, 2, and 3

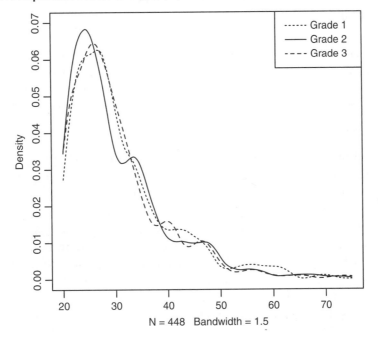

Back-to-Back Histograms

As well as through side-by-side boxplots and kernel smooths, distributions can be compared through back-to-back histograms. These allow us to compare the proportions scoring at any of the levels identified on the left. Each bar can be assessed against the percentage scale at the bottom of the graph, as shown in Figure 5.18.

Figure 5.18: Parent Hyperactivity Ratings by Sex of Child

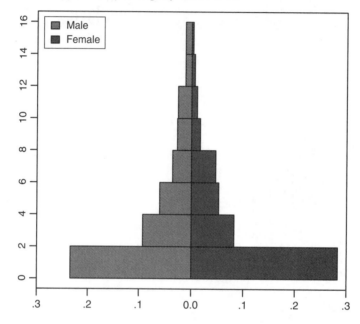

Graphs of this form have often been used to compare the sexes. In Figure 5.18 we see parents' ratings of children on a hyperactivity scale. There are clearly more girls in the lowest category, and more boys at higher levels. Another use of this graphic is to compare treatment and control groups, as has been done in Figure 5.19, using the same ratings as in Figure 5.18.

The Population Pyramid

A variation on the back-to-back histogram is the population pyramid, illustrated in Figure 5.20.

Population pyramids differ from standard back-to-back histograms in that the age groups to be compared are listed down the centre rather than on the left. The bars on each side can still be compared to a percentage scale at the bottom.

These graphs are often used to compare the percentage of the male population who are in a five-year category to the percentage of the female population. Here, the percentages for males exceed those for females at the bottom, but the percentages for females are higher at the top. For both sexes, we see a bulge halfway up the graph, the result of the baby boom after the Second World War. Pyramids of the same form can be used to compare age and sex distributions for different nations or periods in time.

Figure 5.19: Parent Hyperactivity Ratings by Type of Site

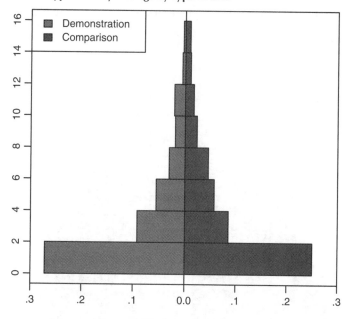

Figure 5.20: Canadian Population Pyramid, 2006

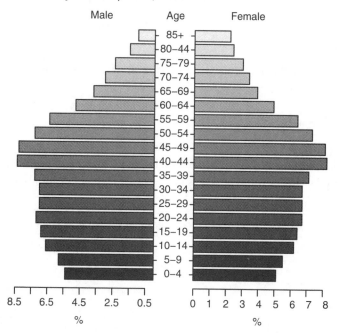

The current Canadian graph is far from a pyramid, but this graphic form was invented at a time of higher birth rates, which expanded the lower bars, and higher death rates, which contracted the upper. Graphs for less economically advanced countries may still have a resemblance to pyramids. For example, a graph for Mexico appears in Figure 5.21.

Figure 5.21: Mexican Population Pyramid, 2010

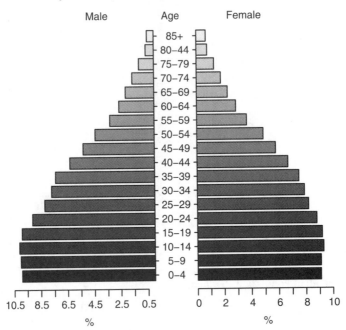

Although Mexican birth rates have declined recently, creating a narrowing at the base, the rest of the graph is roughly pyramidal.

Cumulative Distributions

Although graphs showing cumulative distributions are not routinely seen in social science, it is good to be able to read them. They represent the proportion of cases with scores at or below specified levels. Statistical packages may show percentage like those in Table 5.1.

Table 5.1: An Example of Percentages and Cumulative Percentages		
Score	%	Cumulative %
1	16.7	16.7
2	22.4	39.1
3	35.9	75.0
4	25.0	100.0

The figures in the right-hand column are just the percentages with scores equal to, or less than, those identified on the left.

Cumulative percentages of this kind make up cumulative distributions, which may be graphed as shown in Figure 5.22.

Figure 5.22: Cumulative Distribution of Annual Household Income

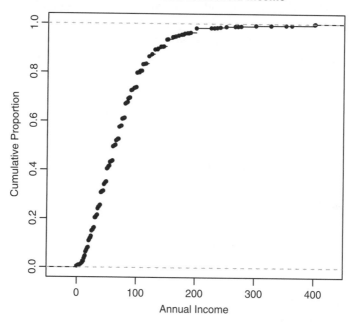

There are far more cases toward the left, so the cumulative distribution function (CDF) rises most rapidly on the left. It begins at zero, at the point at which no cases have yet appeared, and finishes at 1.0, when all cases have been accounted for. Its relatively rapid rise on the left, and slow rise on the right, reflects the rightward skew in these data.

Creating a Mean and Standard Deviation Table

In many published reports you will see a table of means and standard deviations. Such tables conveniently summarize key characteristics of interval and ratio variables, of dichotomies, and of ordinal variables the authors have treated as interval.

The tables are readily created. First we need to choose a monospace (or fixed-width) font, one in which each character is the same width. Numbers must be aligned, and alignment is extremely difficult if characters are of different widths. Common monospace fonts include Andale Mono, Bitstream Vera Sans Mono, Courier, Courier New, DejaVu Sans Mono, FreeMono, Lucida Console, and Monaco. We can choose from among those available to us on grounds of clarity and appearance.

The table heading is sometimes simply "Descriptive Statistics," and that will be fine for this illustration. The heading should be in bold, or perhaps in italics. It must be centred over the columns of the table. If we have not worked out the width of the table, we will have to key in the title and move it later. Here we know the width, so we can centre the title correctly at the beginning.

Table 5.2: Descriptive Statistics

Beneath the table heading, there must be column headings, to indicate which columns contain the variable names, means, and standard deviations. Column headings must be bolded, or italicized, to match the table heading. The heading for variable names must be left justified, and the headings for means and SDs must be centred over the data that will fall beneath them. Here we know where the data should lie, so we can centre the column headings correctly at the beginning.

Table 5.2: Descriptive Statistics

Variable	**Mean**	**SD**

Next we enter variable names. Although a study may involve dozens, to illustrate we need only a few, which will be taken from Nakhaie and Arnold (2010). Variable names are entered without bolding or italicizing.

Table 5.2: Descriptive Statistics

Variable	**Mean**	**SD**
Age		
Change in health		
Religious service attendance		
Perceived love		

Finally, we enter the means and standard deviations. In sociology they are usually reported to three decimal digits. You will see fewer in some other disciplines. Decimal points are always vertically aligned.

Table 5.2: Descriptive Statistics

Variable	**Mean**	**SD**
Age	48.685	15.963
Change in health	−.042	.183
Religious service attendance	3.175	1.268
Perceived love	.961	.193

This form of summary table does not include measures of skewness or kurtosis. Although they could be added, they are usually omitted. Shapes of distributions, where they are important, are often discussed in the text.

Summary

As we have seen, the five-number summary allows us to quickly present key information on a distribution, and the mean and standard deviation table efficiently presents measures of central tendency and dispersion. This chapter has largely been focused on how graphs can complement or extend what we can present with a set of numbers.

The boxplot presents the information found in a five-number summary, but also displays outliers. It makes it easy to see whether a distribution is skewed, and whether it has long or short tails.

Sometimes, though, we need to see what proportions of the data are found at particular values. We may then choose among pie charts, horizontal and vertical bar charts (or histograms), and dot charts. If we need to show only the outlines of a distribution, we can employ smoothed histograms. Or, if we need to show the proportion of cases at or below a specific value, we can use a cumulative distribution chart.

Distributions can be compared though boxplots, back-to-back histograms, or their variation, the population pyramid.

The range of options is broad, so that in a situation where a message might be conveyed graphically or numerically, and when our audience is not familiar with statistics, we are likely to have a live option before us.

Review Questions on Summarizing a Distribution

1. What is a five-number summary? What is its great advantage?

2. Twelve students' grades were as follows:

   84, 83, 82, 81, 78, 78, 78, 77, 76, 74, 72, 69

   Give the five-number summary. Explain how you got the three central numbers.

3. For the data in question 2, draw a boxplot. Explain how you got the numbers for the ends and centre of the box.

4. What happens to a boxplot when two quartile values are the same? How can this come about?

5. How may our description of a distribution vary as a result of decisions made in graphing?

6. Why might we prefer a histogram (or bar chart) to a pie chart? A horizontal histogram (or bar chart) to a vertical?

7. What is an ASH? How can these be useful?

8. What is a kernel smooth? How can these be useful?

9. What are some advantages of a back-to-back histogram? Of side-by-side boxplots for groups?

10. What is the difference between a population pyramid and a standard back-to-back histogram?

Part II
STATISTICAL INFERENCE

Traditionally, the major distinction within statistics has been that between descriptive and inferential statistics. The latter concerns itself with how we ought to judge hypotheses, and what a sample may tell us about the population from which it is drawn. These questions are addressed in Part II.

Chapter 6 deals with sampling distributions. We need to be clear on how results vary from sample to sample if we want to know how well we can estimate from a given sample. Knowledge of sample-to-sample variation is needed to distinguish meaningful differences or correlations from random fluctuations. Sampling distributions, which enable us to estimate how far results will move about from sample to sample, are thus central to statistical inference. In the social sciences, we rely very heavily on four distributions (the normal, t, chi-square, and F). The characteristics of the first three, and how each is used, will be illustrated. (The F distribution will appear later, in the discussion of regression and ANOVA, in which it is used.) The impact of sheer sample size on the accuracy of our findings will also be noted.

Chapter 7 reviews the standard model of inference. The review includes the central ideas of null and research hypotheses, confidence intervals, Type I and Type II errors, and statistical power. Chapter 8 reviews an increasingly used alternative version of inference, the Bayesian model. Alternative versions of probability, including the one on which Bayesian analysis rests, are reviewed. How we can use Bayes' theorem to combine prior evidence (or belief) with fresh data to arrive at a revised picture of the world is illustrated. Bayesian analogues to standard inferential methods are presented (i.e., credible intervals instead of confidence intervals and Bayesian hypothesis testing).

6

Sampling Distributions

Learning Objectives

In this chapter, you will learn

- what a (theoretical) sampling distribution and an empirical sampling distribution are;

- how we obtain empirical sampling distributions, and how they help us;

- the characteristics of three of the most widely useful sampling distributions, and when they are useful; and

- the effects of sample size on sampling distributions.

If we draw samples and calculate the same statistic, perhaps a mean, we will get different answers from one sample to the next. If we store the results for a great many samples, we can view their distribution. This will be an "empirical sampling distribution," which will approximate the "theoretical sampling distribution," commonly referred to as just "the sampling distribution" of the statistic. By definition, this is the distribution of results we will get if we calculate a statistic for an infinite number of samples.

We cannot, of course, draw an infinite number of samples, but if we want an empirical sampling distribution, we have the computing power to do a very great many. Here, for example, are two empirical sampling distributions, based on 100,000 samples. The first graph is based on samples of 100, and the second on samples of 400. In the population from which they were drawn, half the cases were scored 0 and half 1. Let us see how closely the proportion of 1s has been estimated. Figure 6.1 presents the results.

Figure 6.1: Two Empirical Sampling Distributions

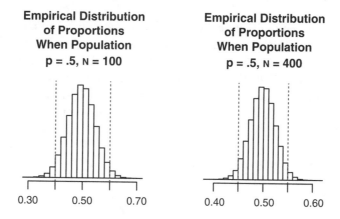

In each graph, the vertical lines divide the central 95% of sample proportions from the outer 5%.

Unsurprisingly, the sample results lie closer to the true proportion of .5 for samples of 400 than for samples of 100. For samples of 100, the range enclosing the central 95% of samples is .60 − .40 = .20. With samples of 400 the central 95% are found in a range half as wide (.55 − .45 = .10). There was no reason to expect we would overestimate more often than underestimate, so it is again unsurprising that the graphs are close to symmetric. We may also notice that the graphs resemble the normal distribution, as, on theoretical grounds, they should.

Fortunately, we do not need to do simulations routinely in social science, because the theoretical sampling distributions for the statistics we typically use have been worked out. Simulations may still be helpful in two basic situations. In the first, the theoretical results are "asymptotic"—that is, they tell us the shape of the distribution we will get as sample sizes tend to infinity. Such results leave us uncertain how many cases we need for the distribution to be well approximated. Fortunately, the needed simulations have been done for many common statistics, so we can often use established rules of thumb about necessary sample sizes.

Simulation is also an option when we do not have a theoretical distribution. For example, we know that the sampling distribution for a median, if we draw large enough samples from a normal distribution, will approximate the normal. In other situations, it is often difficult to work out the theoretical distribution. Dealing with such situations is more the province of statisticians than social scientists, but we can use simulation if we need to.

However obtained, sampling distributions are central to statistical inference. If we know how results vary across samples, we can calculate how likely it is that our results will depart from the true figure by a given amount. We can also calculate how often we will get a figure in the tail of the distribution through random variation across samples.

Fortunately, most sampling distributions we work with in the social sciences follow one of only four forms: the normal, the t, the chi-square, and the F distributions. The first three will be dealt with in this chapter, and the F distribution will appear in Chapter 14, after the section on Analysis of Variance.

The Normal Distribution

In the terms employed to describe distributions, the normal is continuous, unimodal, and symmetric. The standard normal distribution, with mean 0 and SD 1, is by definition mesokurtic. Normals with other means and SDs, when graphed on the same axes, may appear rather different. To distinguish among them, we refer to their means and SDs. For example, one with a mean of −1 and an SD of 2 is said to be N(−1,2), where the N means that it is a normal distribution, and the figures in brackets are the mean and SD. Three different normals, with the standard distribution in the centre, are shown in Figure 6.2.

Figure 6.2: Three Normals: (−1,2), (0,1), and (1.8,.6)

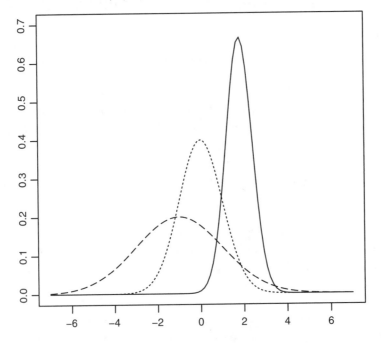

By the Central Limit Theorem, the sampling distribution of the mean for any variable with a definable SD will be normal if the sample size is great enough. To see what happens for a given N (number of cases), we can draw many samples of that size and save the results. For examples, see the graphs in Figures 6.3 and 6.4. The first set shows four theoretical distributions. The second set shows the distribution of the means for samples of 100 from the same theoretical distributions. Despite the very different distributions from which cases were drawn, the means approximate the normal distribution.

Another set of graphs, found in Figure 6.5, illustrates a further point: means for samples from symmetric distributions may approximate normality rather well with Ns of 30 (as shown in the upper two graphs), but we often need larger samples when drawing cases from skewed distributions, as illustrated in the lower two graphs, which are slightly skewed.

Figure 6.3: Four Distributions from Which Cases Have Been Drawn

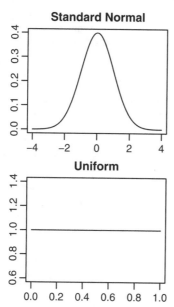

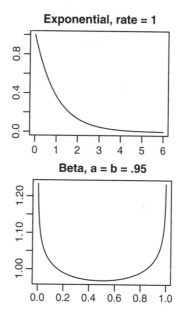

Figure 6.4: Four Sampling Distributions

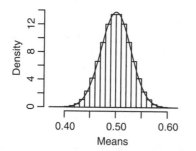

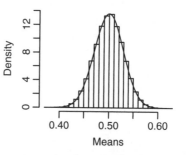

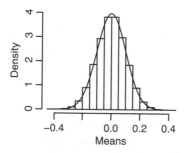

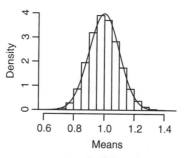

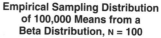

Figure 6.5: Four Sampling Distributions, N=30

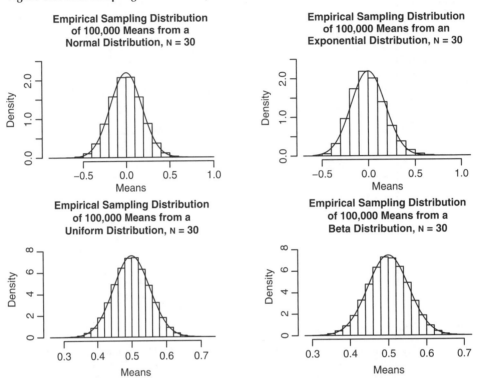

If a sampling distribution is close to normal, we can take advantage of a key feature of the distribution: that 68% of the cases lie within one SD of the mean, and 95% within 1.96 SDs. (As an approximation, people often say within two SDs.) From this, we can work out the fraction of samples that will lie within a specific range.

For moderate Ns, the sampling distribution of a mean follows the normal curve to a good approximation. You will see different rules of thumb as to when the normal curve is adequate: 25, 30, 50, and 100 have all been suggested. How many cases will be needed will depend on the skewness of the distribution from which we are sampling, and the degree of precision required.

As we shall see in Chapter 9, moderate sample sizes lead to approximately normal sampling distributions for many measures of association.

To avoid confusion between the SD of a variable and that of a sampling distribution, we call the SD of a sampling distribution the standard error. Using this term, if a sampling distribution is normal, 68% of samples will lie within one standard error of the mean, and

95% within 1.96 (or, as an approximation, within 2). If our estimate is unbiased, the mean of the sampling distribution will be the true population figure. Thus, 68% of samples will lie within one standard error of the true figure and 95% within 1.96.

For example, suppose we want to take a random sample of 100 students in a second-year course in social statistics, who will be asked to respond to a scale for math phobia. Suppose the SD for the scale is known to be 20, and we wish to estimate the standard error (SE) of the mean before drawing the sample. The basic formula is

$$SE(x) = SD(x) / [N]^{.5},$$

where SD(x) is just the standard deviation of the variable of interest, and N is the sample size.[1] Substituting, we obtain

$$SE(x) = 20 / [100]^{.5} = 20 / 10 = 2.000$$

On this basis, we can say that 68 sample means out of 100 should lie within 2.000 scale units of the true population mean, and

95 of 100 within approximately 4.000 scale units of the true figure $[2(2.000) = 4.000]$

The Sampling Distribution of a Proportion

A proportion can be seen as the mean for a variable coded 0 and 1, so we can get the standard error for a proportion by placing the standard deviation of a proportion in the numerator. An explanation of the formula for the SD is found in an appendix to this section. Denoting the proportion of 1s as p, the formula is

$$[p(1 - p)]^{.5}$$

so the standard error becomes

$$SE(p) = SD(p) / N^{.5} = [p(1 - p)]^{.5} / N^{.5}$$
$$= [p(1 - p) / N]^{.5}$$

Suppose p = .2, and N = 400. Then,

$$SE(p) = [.2(.8) / 400]^{.5} = [.16 / 400]^{.5} = .4 / 20 = .2 / 10 = .02$$

For the normal approximation to the sampling distribution to be acceptable, a common rule of thumb is that Np and N(1 − p) should both exceed 10. The distribution of results for three different values of Np is shown in Figure 6.6. The distribution is clearly skewed to the right for Np = 5. The skew implies that large overestimates will come up more often than large underestimates. With Np = 10 skew remains readily visible, although less pronounced, and at 20 it is still less pronounced. (For these values, Sk declines from .413 to .267 to .150.)

Figure 6.6: Empirical Sampling Distributions for Three Values of Np

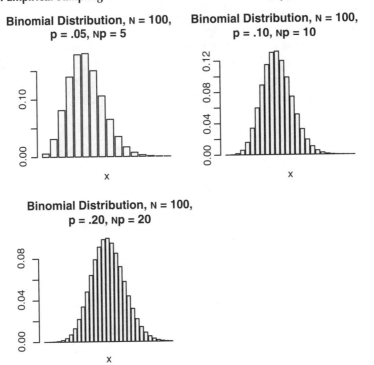

The t Distribution

When we have small samples, or we wish to be very precise about the sampling error for a mean obtained from a sample, we rely on the t distribution. Just as there are many normal distributions, so there is a whole family of "t distributions," but the convention is to refer to them as a single distribution. The members of the t family have several important features in common:

1. like the normal, they are symmetric and unimodal;
2. compared to the normal, they have low central peaks and heavy tails; and
3. as sample size rises, they become more and more like the normal: their central peaks rise, and their tails lighten.

These points are illustrated in the graphs in Figures 6.6 (above) and 6.7 (below).

The t distributions shown are distinguished by DF (degrees of freedom). If we are estimating means, DF = N − 1, so 2 DF implies an N of 3, and so on.

Another way to see the difference between t distributions with varying DF and the standard normal is to focus on their tails, as shown in Figure 6.8.

Figure 6.7: Three t Distributions and a Normal

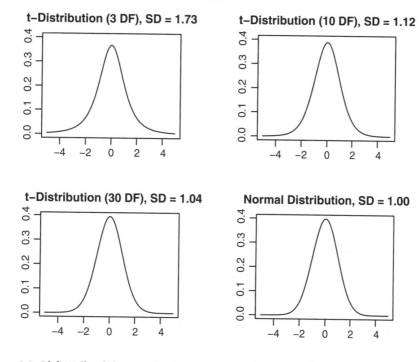

Figure 6.8: Right Tails of Three t Distributions and the Normal

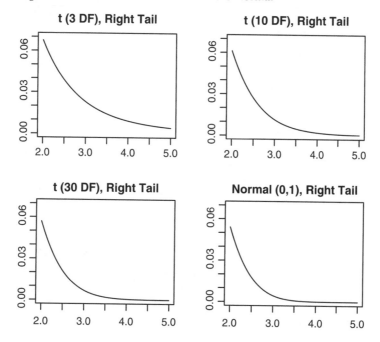

Since we will encounter degrees of freedom again later, we should be clear on how they are defined: the number of DF equals the number of data values which are free to vary.

Let us suppose that we work out the mean for five numbers: 1, 3, 4, 5, and 7. It comes to $(1 + 3 + 4 + 5 + 7) = 20 / 5 = 4$. We would get the same mean for any five numbers summing to 20, but a different mean for a different sum.

According to the formula, DF = N − 1, we have $5 − 1 = 4$ DF. To see that this is so, we observe that the total for our first four scores is $1 + 3 + 4 + 5 = 13$. We have a mean of 4, which we obtained only because the overall sum of five values was 20, so the fifth value had to be a 7. Given the mean, the final score is not free to vary, although the first four are. They might, for example, have been 1, 2, 3, and 4, adding up to 10. If they were, though, the fifth value would have had to be 10 to preserve a total of 20 and a mean of 4. Essentially, once we have the mean, the first N − 1 scores can move freely, but the Nth must have a specific value for the mean to come out correctly.

The Sampling Distribution for the Mean When t Applies

Because the tail of a t distribution is heavier than that of a normal distribution, we have to go more standard errors away from the mean to pick up 95% of the cases. The smaller the sample becomes the farther we have to go.

To illustrate the difference moving to a t distribution makes, let us look at what happens with a sample of 16 cases. Above we assumed that the scale of mathematics phobia had a

Figure 6.9: Distance from Mean Encompassing 95% of Samples, n = 100, DF = 3–30

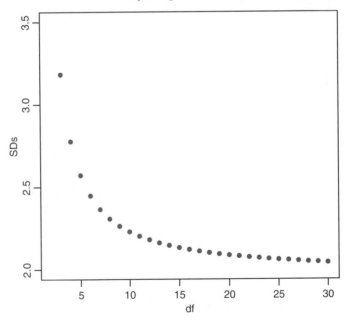

mean of 20 and a standard deviation of 20. With an intended sample size of 16, the standard error of the sampling distribution would be

$$\text{SE}(x) = \text{SD}(x) / [\text{N}]^{.5} = 20 / [16]^{.5} = 20 / 4 = 5$$

Reducing our sample size from 100 to 16 has increased our standard error by a factor of 2.5. Further, to encompass 95% of samples, we must multiply the standard error not by 2, but by 2.12. Thus 95% of samples of 16 will lie within 10.6 scale units of the true population figure [2.12(5) = 10.6].

As the formula indicates, when sample size declines, our standard errors increase with the inverse square root of sample size. However, our margin of error increases further because we must work from a sampling distribution whose tails become heavier as N declines.[2] To illustrate, Figure 6.9 shows us the number of SDs we must go from the true population mean to encompass 95% of samples when DF ranges from 3 to 30.

The Chi-Square Distribution

As with t, so with chi-square, we ordinarily refer to "the chi-square distribution" when in fact we have a family of distributions. Another oddity about the terminology is that, by tradition, we speak of chi-square, not chi-squared, even though the distribution is based on squared quantities.

To denote the theoretical chi-square distribution, we typically use the upper case Greek chi:

$$\chi^2$$

As we have seen earlier, in referring to a calculated statistic which should follow the χ^2 distribution, we typically use the upper case Roman X, which resembles the Greek chi.

The simplest chi-square distribution is based on a normal distribution with mean 0 and SD 1. In this distribution, every case is referred to as a normal deviate. To obtain a chi-square distribution, we square the value of each deviate. Since all the values are squared, the shape, central tendency, and dispersion of the resulting distribution will differ sharply from those of the original. As before, the distribution will be unimodal, with its peak at the value of 0, but

1. the distribution will be right skewed, rather than symmetric;
2. because there are no longer values below 0, the mean must be greater than 0; in fact it is 1.0; and
3. because extreme values are squared, the SD increases; in fact, it turns out to be $2^{.5}$.

Suppose we were to draw two normal deviates at random, square their values, and sum them. Suppose we were to do this many many times. How would the resulting sums be distributed? The answer is that they would approximate a second of the chi-square distributions. If we were to draw three normal deviates, and square and sum them many many times, we would approximate another chi-square distribution.

Technically, the simplest chi-square distribution is the sampling distribution of a single squared normal deviate. The second chi-square distribution is the distribution of the sum of two squared normal deviates, and so on. Since it is awkward to refer to "the sum of n squared normal deviates," we refer to chi-square distributions in terms of degrees of freedom (DF). Each DF refers to a squared normal deviate, so the simplest chi-square distribution has one DF, the next has two DF, and so on. In a formula,

$$X^2 = \sum z^2,$$

where the number of z's to sum equals DF.

As DF increase, the shape, central tendency, and dispersion of the chi-square distribution shift. The mean increases by 1.0 for each DF, thus always equalling DF. The standard deviation rises more slowly, always remaining equal to $[2DF]^{.5}$. Gradually, the distribution becomes more symmetric. The changes that occur are illustrated in Figure 6.10.

Figure 6.10: Four Chi-Square Distributions

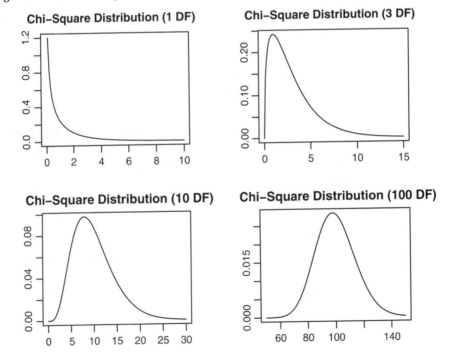

The chi-square distribution is often used in dealing with two-way crosstabulations. For a table of this kind, we calculate chi-square as

$$X^2 = \Sigma \, (O - E)^2 \, / \, E \, ,$$

where the Os are the observed values in the cells, and the Es are the values expected on the assumption that rows and columns are independent. The summation sign reminds us that we have to sum results across all the cells in the table. To remind ourselves how this works without being distracted by details of computation, let us consider the following fictitious table.

	C1	C2	Total
R1	15	5	20
R2	5	15	20
Total	20	20	40

We obtain expected cell values on the assumption that the row and column variables are independent from a standard formula:

$$R(i)*C(j) \, / \, N \, ,$$

where $R(i)$ is the row total for row i,

$C(j)$ is the column total for column j, and

N is the table total.

$R(i)*C(j)$ / N gives us, for cell (1,1)

$$20*20 \, / \, 40 = 10$$

Since the marginal totals for the rows and columns are the same, we will get the same expected values for the other cells.

To get X^2, we first take the difference between O and E for each cell. Next we square the difference, and divide it by E. Finally, we sum the result across cells. These steps can be carried out in a worksheet, as shown in Table 6.1.

Table 6.1: A Worksheet to Calculate Chi-Square					
Cell	O	E	$(O - E)$	$(O - E)^2$	$(O - E)^2 / E$
(1,1)	15	10	5	25	2.5
(1,2)	5	10	−5	25	2.5
(2,1)	5	10	−5	25	2.5
(2,2)	15	10	5	25	2.5
				$X^2 = \Sigma \, (O - E)^2 / E =$	10.0

X^2 was invented to test whether row and column variables are independent. If its value is high enough, we abandon the idea that they are. How we do this will be reviewed in Chapter 7. Here, we just need to understand a point about its sampling distribution.

Since for samples of adequate size, this distribution is X^2, we can compare the figure calculated from the formula with the X^2 distribution to see how readily our result might have arisen by chance. To do so, we need the DF for the table, obtained from the formula

$$DF = (r - 1)(c - 1) \,,$$

where r is the number of rows in the table and c the number of columns. For a 2 × 2 table, $DF = (2 - 1)(2 - 1) = (1)(1) = 1$. For 1 DF, the .05 critical value is 3.84. That is, the value below which we expect to find X^2 in 95 samples out of 100, if rows and columns are truly independent, and we have sampled randomly, is 3.84.

In practical work social scientists do not often need to remember critical values, since statistical packages will provide the p-values that correspond to specific values of X^2. It is good, though, to understand why we say that a table has a given number of degrees of freedom. Suppose we have a 2 × 2 table, with fixed marginals, like the following:

		10
		10
10	10	20

Suppose we fill in one of the cells:

5		10
		10
10	10	20

From this cell value and the row total, we can fill in the value for the other cell in the row.

5	5	10
		10
10	10	20

Similarly, having one entry in each column, and the column totals, we can fill in the other cell in each column.

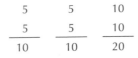

5	5	10
5	5	10
10	10	20

In a 2 × 2 table, given the marginals, only one cell value is free to vary. When we have fixed the value of any cell, all the others are determined. In this sense, we have only one degree of freedom.

In larger tables, the number of DF also equals the number of cell values free to vary. For a 3×3 table, for example, $(r - 1)(c - 1)$ yields $(3 - 1)(3 - 1) = (2)(2) = 4$ DF. If we have a 3×3 table with fixed marginals, and fill in the value of 4 cells, as has been done below, all other cell values can be determined in the same way as those in the 2×2 table we have just seen.

5	5		15
5	5		15
			15
15	15	15	45

The mathematics to show that the number of cell values free to vary is linked to the same number of squared normal distributions is outside the range of this text. Here we just need to understand how the expression "degrees of freedom" is used in each case, and to recognize that when we have a certain number of DF in a table, we test for the independence of rows and columns using the chi-square distribution with equivalent DF.

Relations among the Normal, t, and Chi-Square Distributions

We have seen that the t distribution approaches the normal as DF increase, and that the chi-square distribution is created from the sum of squared normal distributions. There is another connection among the three: the t distribution results from dividing a normal distribution by a chi-square. Formally,

$$t = N(\mu,\sigma) / \chi^2(DF) ,$$

where μ is the mean of the normal distribution, and σ is its SD. The various t distributions differ in the DF of the chi-square in the denominator.

The Effect of Sample Size

For all distributions ordinarily used by social scientists, standard errors decline as sample sizes rise. This decline in standard errors is linked to the Law of Large Numbers.[3] It holds that as samples grow larger statistics calculated for them tend toward the results we would obtain if we had data on the full population. To illustrate, we shall look at means from samples of increasing size.

Specifically, these will be running means, means calculated from all the cases drawn up to a given point, which change gradually as further cases are drawn. We shall examine running means based on cases from a standard normal distribution, others based on cases from a t distribution with three DF and others from a chi-square distribution, again with three DF. For the normal and the t distribution, we can expect the running means to converge toward the distribution means of zero. As pointed out above, the mean of a chi-square distribution equals its DF, so the running means should converge to 3. The results are shown in Figure 6.11.

Figure 6.11: Running Means for Samples from Three Distributions

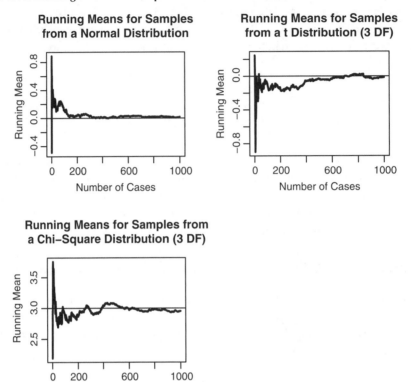

In each case, the initial means are some distance from the values obtained with greater sample sizes. In each case, though, the running means have settled down near the expected values for their distributions, although the normal, with its relatively light tails, has settled down more quickly than the others.

Summary

If we do not know how results are likely to differ from one sample to another, we can neither say how accurate our estimates are nor test hypotheses against the possibility that results came up by chance. If we know the sampling distribution of the statistics we are using, then these assessments become possible. In the social sciences, we rely largely on four sampling distributions, three of them covered in this chapter: the normal, the t, and the chi-square distributions. When none of these works well, we can draw random data to create an empirical sampling distribution. Ordinarily, though, we do not need to.

The sampling distribution of the mean or of a proportion (which is a kind of mean) tends to approximate the normal for moderately sized samples. For smaller samples, the t distribution serves us well. When we are working with crosstabulations, we often use the chi-square distribution. Relying on these distributions, software provides us with estimates of how accurate our results are likely to be, and provides tests of hypotheses. How they are tested will be covered in Chapter 7.

Review Questions on Sampling Distributions

1. What is a (theoretical) sampling distribution? An empirical sampling distribution?

2. What is meant when we say that a sampling distribution has been estimated by simulation?

3. What is a standard error?

4. What does the Central Limit Theorem tell us?

5. What is one statistic for which the sampling distribution is often hard to work out analytically?

6. For each of the following statistics, give the appropriate sampling distribution.

   • mean, from small sample _____

   • mean, from large sample _____

   • test of independence for a crosstabulation _____

7. What are three ways in which the t distribution resembles the normal?

8. What are two ways in which the t distribution differs from the normal?

9. What is the relationship between the chi-square distribution and the normal?

10. How does the mean of a chi-square distribution with one DF differ from that for a normal distribution? Why?

11. How does the SD of a chi-square distribution with one DF differ from that for a normal distribution? Why?

12. Why does the standard error for a mean become greater as sample size becomes smaller?

13. In what sense are there many t distributions? Many chi-square distributions?

14. For a chi-square, how does the sampling distribution change with DF? Why does this matter?

15. For t, why must we be concerned about degrees of freedom?

16. Suppose someone calculates the standard error of the mean for some variable, and obtains 1.0. Assuming simple random sampling, within what range of the true mean would 95% of sample means lie?

17. Give the formula for the standard error of a proportion. If the true proportion having some characteristic is .50, and we take a simple random sample of 100, what is the standard error?

18. What is a suggested rule of thumb for when the sampling distribution of a proportion can be treated as normal? What did graphs suggest about the accuracy of this rule of thumb?

19. What does the Law of Large Numbers tell us?

Notes

1. If we do not know SD(x), and must estimate it from a sample, an unbiased estimate will be obtained by using (N − 1) rather than N, but the difference will be trivial except for small samples.

2. We are estimating the standard error on the assumption that we know the SD(x). If it had to be estimated from a sample, we would lose one DF. Instead of N in the formula, we would use (N − 1) so the standard error would be a bit larger.

3. Several related laws have been defined, which for our purposes need not be distinguished.

The Standard Model of Statistical Inference

Learning Objectives

In this chapter, you will learn central concepts in the standard model of statistical inference, including

- null hypotheses and research hypotheses;
- one and two-tailed significance tests, with an illustration of a one-tailed test using chi-square;
- statistical power; and
- confidence intervals.

Central Ideas

Social scientists typically use a model of statistical inference derived from Sir Ronald Fisher, Jerzy Neyman, and Egon Pearson. A thriving alternative is presented in Chapter 8, but first we must understand what has been the standard model in our field.

It began with the agricultural experiments of Sir Ronald Fisher, who wanted to be reasonably sure that differences in crops were produced by different farming methods or varieties of seed rather than by random variation, and thus invented the significance test. As extended and formalized by Neyman and Pearson, the standard approach requires us to define

a) a null hypothesis, often labelled H_0, which usually states that

there is no difference between groups, or

there is no association between variables.

Since these hypotheses usually state, in effect, that nothing is happening, it is easy to see why they should be referred to as null hypotheses. However, the name still applies when someone specifies a difference or association which is not zero but which is to be tested against another possibility.

b)   a research hypothesis, often labelled H_1, which typically states that

there is a difference between groups, perhaps with a prediction as to which will score higher, or

there is an association between variables, perhaps expected to be positive or negative.

c)   a test of significance. We must ask whether the observed difference between groups, or the observed association between variables, is likely to have come up by chance. We get our answer through a test of significance. Such a test basically asks how often we would get the results we have, or more extreme results, if the null were true. A test of significance results in a p-value, which provides the probability of getting the result we have, or a more extreme one, by chance. By convention, if the p-value is less than .05, we say our result is significant. We then take it seriously as evidence against the null.

Tests of significance are based on sampling distributions. If we have results in the tail of a sampling distribution, which would be infrequent as a result of sampling variation if the null were true, we will typically take our results seriously. The tail region in which we will declare our results "statistically significant" and take them seriously is known as the "critical region." The graphs in Figure 7.1 show the critical regions for some tests based on the normal sampling distribution, the chi-square, and the F distribution. Each critical region is based on a p-value of .05.

Tests may be one- or two-tailed. In a one-tailed test, results are significant only if they fall in the tail of the distribution specified by the research hypothesis. In a two-tailed test, an extreme result in either tail will be taken as significant. Tests employing the chi-square and F distributions are ordinarily one-tailed. Those based on the normal and t distributions may be either one- or two-tailed, but in contemporary social science are usually two-tailed.

However we set them up, significance tests are subject to two kinds of error:

In Type I errors, we abandon the null when it is true—that is, we take a result seriously when it came up by chance.

In Type II errors, we hold to the null when it is false—that is, we treat a result as having come up by chance when it reflects a real difference or association.

The two types of error can be traded off. If we wish to reduce Type I errors, we can use a p-value below the usual .05 in deciding whether to take a result seriously. If we use .01, then if the null is true our chances of falsely rejecting it are reduced by a factor of five. The trouble is that if we do this the chance of making a Type II error—of treating a result as due to chance when it is not—goes up.

Figure 7.1: Critical Regions for One- and Two-Tailed Tests

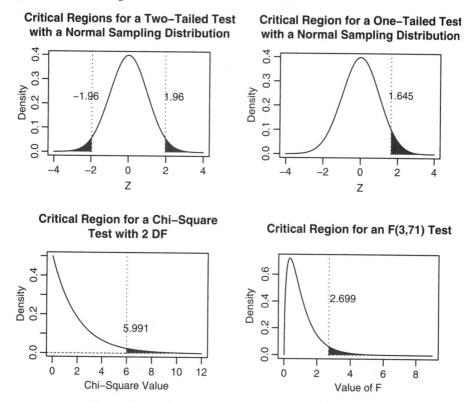

If we wish to reduce Type II errors, we can shift the required p-value upward. If we use .20 as our criterion, as might be done in an exploratory study, then we are much less likely to miss a real difference or association, but if the null is true we are four times more likely to take a result seriously when we should not.

Although by shifting the p-value we can modify the balance of risks, each form is problematic. Taking larger samples, and thereby reducing standard errors, will reduce the size of Type I errors and the probability of Type II errors.

To see what happens when we do a test of significance, let us consider the difference between two means. Earlier we graphed scores on the CESD depression scale, but we have not made a formal test for differences at different times. Let us see whether we can detect a difference in means between parents tested when their children were in Grade 1 and the same parents when their children were in Grade 3. Our null is that there is no difference, and we need to see whether there is evidence to throw it into doubt.

Since we are working with the same cases, the variable of interest is the difference in depression scores. Its mean is −.804—that is, the scores at Grade 3 are a bit smaller. The standard error is .482.[1] We can graph these results as shown in Figure 7.2.

Figure 7.2: A Test of the Null That the Difference in Scores Equals Zero

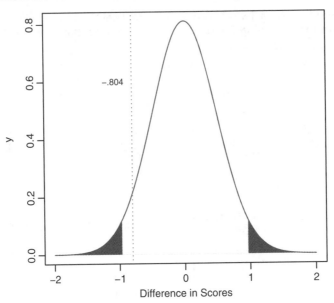

The observed difference is approaching the critical region, but does not lie inside it, so we are left holding to the null.

In practice, for a test of this kind we would not typically get a graph in our output. We would get the two means, their difference, the standard error of the difference, and a p-value, and perhaps some other things. In this case, p would be .095. This would mean that our result lies at the edge of a region in which 9.5% of sample differences would lie if there had been no change in the respondents.

If we were doing two-tailed tests for means or measures of association whose sampling distributions are the normal or the t, a graph, if presented, would resemble the one above. Ordinarily, though, we would not see a graph, but rather the value of the measure, its standard error, the ratio of the measure to its standard error (often labelled t, rather than x / SE(x)), and the associated p-value. For example, results for a mean could look like this:

mean(x)	SE(x)	t	p
.473	.110	4.294	.000

Tests with Chi-Square

Working with chi-square, the test is almost always one-tailed. We have seen above that chi-square itself is obtained from

$$\Sigma\,(O - E)^2 / E$$

The idea behind chi-square is to see how far from independence the observed values lie. That being so, the null hypothesis tested with chi-square, for a crosstabulation, is typically that there is no association between the row and column variables. In different words, the null is that the rows and columns are independent. From the formula, it is clear that for each cell in the table we compare the observed value (O) and the value we would expect to get if the row and column variables were unrelated (E). We square the differences, then sum them (having divided in each case by E). Because we sum a function of the differences between observed values and those expected under independence, a high value chi-square is evidence against independence, and against the null hypothesis.

We have seen above that the shape of a chi-square distribution depends on its degrees of freedom. This means, in turn, that how far to the right we must move to reach the critical region depends on DF. Fortunately, software will deal with this question for us.

To see how the test works, consider Table 7.1, showing smoking status by parental status.

Table 7.1: Smoking Status by Parental Status

	Single Parent	**Partnered Parent**	**All Cases**
Smoker	82	126	208
Non-smoker	103	397	500
Total	185	523	708

To get the Es, we calculate R(i)*C(j) / N for each cell. An appendix provides the rationale for the formula, and Table 7.2 shows the Expected Values.

Table 7.2: Expected Values for Smoking Status by Parental Status

Cell	R(i)	C(j)	N	R(i)*C(j) / N
(1,1)	208	185	708	54.35
(1,2)	208	523	708	130.65
(2,1)	500	185	708	153.65
(2,2)	500	523	708	369.35

Having the expected values, we can set up a worksheet, as in Table 7.3.

Table 7.3: A Worksheet to Calculate Chi-Square for Smoking Status by Parental Status

Cell	O	E	(O – E)	(O – E)²	(O – E)² / E
(1,1)	82	54.35	−27.65	749.52	14.07
(1,2)	103	130.65	27.657	49.52	5.85
(2,1)	126	153.65	−27.65	749.52	4.98
(2,2)	397	369.35	27.657	49.52	2.07
				$X^2 = \sum (O - E)^2 / E =$	26.97

For a 2 × 2 table, we always have 1 DF. (From the formula $(r − 1)(c − 1) = (2 − 1)(2 − 1) =$ $1*1 = 1$.) For 1 DF, the .05 critical value lies at 3.84, and we are far from it. In fact, this result is far into the extreme tail of the sampling distribution, as shown in Figure 7.3.

Figure 7.3: Tests of the Null that Smoking and Parental Status Are Independent

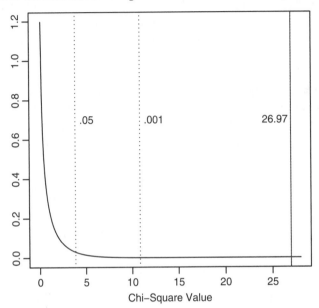

This graph has no shaded critical region. The region is too small to be shown, since the value of chi-square, shown on the right, is so far into the tail. The .05 and .01 critical values, shown to its left, have been left far behind.

In practice, we would not ordinarily see a graph of this kind. Software would provide the value of chi-square, the DF and the p. A report would say something like

$$\text{Chi-square(1 DF)} = 26.97 \text{ , p} = .000 \text{ ,}$$

telling us that the probability of getting a result so extreme, or more extreme, by chance, with 1 DF, is too small to register within three decimal digits.

Since in practice software provides our results, and nothing new conceptually is required to deal with tables of larger dimensions, we need not consider them here.

Statistical Power

When a result is not statistically significant, a possible reason is that the sample is too small to pick up something real. Before we examine the data, or before we even gather it, we can work out how likely we are to get a significant result if an effect of a given size is present. (This assumes that we are using some form of random sampling.) Our ability to detect an effect of a given size is called the power of the test. It equals 1 minus the probability of missing an association that is real or in different terms, of making a Type II error. More formally,

$$\text{Power} = 1 − \text{p(Type II error)}$$

Suppose that, to take a figure from a previous example, there was a difference of −.804, the size we have observed, in the population, and that the standard deviation of the differences was, as in the sample, 10.433. What would be the probability that a sample of the size we have would yield a significant result? We typically rely on software to answer questions like this, so no calculations will be shown, but the answer turns out to be .370. That is to say, if there were a difference of this order, 37% of samples would be expected to yield a result significant at .05.

In designing a study, we can get the probability of detecting effects of different sizes with different Ns. Figure 7.4 shows what sample sizes would be likely (and unlikely) to let us detect an effect of the size we have seen and one twice as large.

Because sample size is so important to detecting effects, some research funders require estimates of the power to detect effects of given sizes. Some, for example, want the probability of detecting an effect of a size deemed plausible to be .80. In the graphed example, 1,000 cases would be insufficient for the smaller effect, whereas 400 would be fine for the larger.

Figure 7.4: Power to Detect Effects

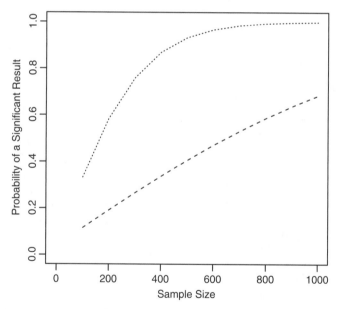

Numerous writers have pointed out that, since larger and larger samples allow us to reach statistical significance for smaller and smaller effects, we ought not to treat a result as important simply because it is statistically significant. A more skeptical view holds that any predictor we reasonably think should have an effect is unlikely to do exactly nothing. If so, a hypothesis test only tells us whether the sample is large enough to detect the effect, not whether there is one! However we assess this view, we do need to recognize that large samples may yield statistically significant but trivial findings.

Confidence Intervals

Often it has been suggested that, along with p-values, researchers should report confidence intervals. These are simply ranges within which a given proportion of sample results can be expected to fall. A 95% confidence interval is one in which 95% of sample results can be expected to fall, and a 99% interval should cover the results of 99% of samples.

Of course, to say where sample results should lie requires us to know the true population figure. In practice, under the standard approach, we treat our sample result as if correct. To construct a confidence interval we start from our observed result, and then place bounds around it. If the sampling distribution is normal, then to get a 95% interval we go 1.96 standard errors on each side of our observed result. For broader or narrower intervals we simply go more or fewer standard errors from it.

In Figure 7.5 we graph mean monthly income for four cultural groups: anglophone (labelled "anglo"), francophone (labelled "franco"), Native, and other. The vertical bars mark off the tops and bottoms of the confidence intervals.

Figure 7.5: Monthly Income by Ethnicity, with Confidence Intervals

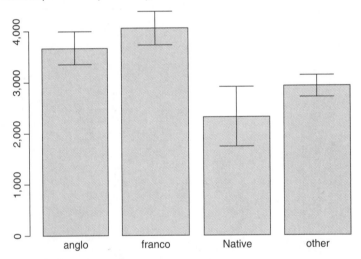

The width of the intervals varies. That for the Native population is largest, because the sample is smallest. The interval for the "other" category is smallest, not because of a large sample size, but because there is least variability in incomes for that category, and thus a small standard error. (Recall that for a mean the standard error is given by $SD(x) / N^{.5}$).

We may also see confidence bands placed around trend lines. In Figure 7.6 we have a plot showing the effect of years of education on current salary, with a 90% confidence band.

The band reflects sampling variation in the mean of both the independent and the dependent variable as well as in the slope. Typically confidence bands are wider for low and high values of the predictor than for values near its mean. The line is algebraically constrained to go through the centroid, so modest variations in the slope will leave the line in roughly the same place when the predictor is near its mean. The same variation in slopes will have more visible effects at the extremes.

Figure 7.6: Years of Education Effect Plot

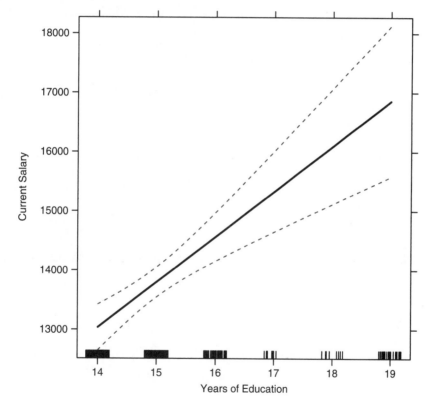

Summary

In this chapter, we have reviewed the central ideas of the standard model of statistical inference: null and research hypotheses, one- and two-tailed tests of significance, Type I and Type II errors, statistical power, and confidence intervals. We have seen how graphs can be used to present the tests and confidence intervals, and also how in practice results are apt to be supplied in output.

Although the standard model of inference is used widely in social statistics, there is an alternative logic of inference, which allows us to deal with some problems which cannot be handled under the standard model. Chapter 8 introduces Bayesian inference.

Review Questions on the Standard Model of Statistical Inference

1. Typically, what is the difference between what a null hypothesis states and what a research hypothesis states?

2. What is a two-tailed test? A one-tailed test? What are some situations in which we tend to use each?

3. What is a Type I error? How can we reduce the chances of one?

4. What is a Type II error? How can we reduce the chances of one?

5. What is the difference between a research hypothesis and a null hypothesis?

6. What is statistical power? Why is it important?

7. What is a confidence interval? How are they constructed?

8. In graphing regression results, why are confidence bands wider at the top and bottom than near the centroid?

Note

1. The SD of the difference score is 10.4332, and N is 448. From the formula $SD(x) / N^{.5}$, we obtain $10.433 / 21.660 = .482$. Taking the difference over its standard error gives us $-.804 / .482 = -1.668$.

The Bayesian Alternative

Learning Objectives

In this chapter, you will learn

- how, in the Bayesian approach, we can combine initial knowledge with data to revise our picture of the world;

- the conception of probability underlying the Bayesian approach;

- Bayes' theorem, on which this form of inference is based;

- how the probability that a specific outcome will take place is calculated; and

- credible intervals, which tell us the range within which a true figure for a population lies.

Although well known in statistics and econometrics, and increasingly used in other fields, the Bayesian approach is not as well known among social scientists, apart from economists. However, many statisticians prefer the alternative Bayesian approach and probably most use both. There are now full textbooks designed to teach social scientists how to use Bayesian statistics. Here, we shall go through its central concepts and methods.

In the standard approach, as summarized in Chapter 7, we begin by stating a null hypothesis: usually that there is no difference between two or more groups, or that there is no association between two or more variables. We then test to see whether to hold to the null. Usually we do not if we obtain results that, if the null were true, would come up by chance less than 5 times in 100.

To get a confidence interval (CI), we obtain the difference between groups, or the association between variables. Then we ask how far the results from other samples might lie from the figure we have. Typically, we seek a range within which they would be expected to lie 95 times out of 100. Since results will vary, so will the bounds of confidence intervals, as shown in Figure 8.1.

Figure 8.1: CIs for Samples of 100 from a Standard Normal Distribution

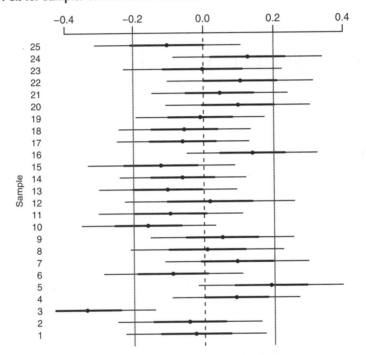

The Bayesian approach aims to answer a different set of questions. It does not use a null hypothesis as such. Instead it tries to clarify how likely it is that the true difference between groups, or the association between variables, is such-and-such, or lies in such-and-such a range.

Confidence intervals do not answer this question. Typically, given a sample, we act as though the estimates we make from it are correct, and then see how much other samples would be likely to vary around them. On Bayesian premises, what we should try to do, given a sample, is to say what the most likely figure is and how far from this figure the truth might plausibly lie. Confidence intervals don't provide a range within which the true figure is likely to lie, but rather a range within which sample results are likely to lie.

To see how likely it is that the true figure is such-and-such, we need to proceed in stages. First, we need to assess what we know before looking at our data, and to try to assign probabilities to each of the outcomes in which we are interested. Then, after gathering data, we can revise our views. To take a relatively simple example, suppose that in some population it is known from large and careful studies that about 6% have an autism spectrum disorder

(ASD). Suppose that you wish to get a better idea of whether a specific child has one. Initially, all you know is that, for children where the family lives, p = .06.

Suppose then that a parent answers the questions in a screening test. It has been designed to pick up almost all of those who would be diagnosed if the child were examined by clinical specialists. We shall suppose that it picks up about 98%. To pick up so many genuine cases, it has to give us some "false positives"—that is, suggest further examination for children who are normal. When children score above the cutoff point suggesting further examination, about 60% have ASD and the others do not (although they may have a different disorder). If 1,000 parents from this population do the test, we can expect the following results.

ASD present	60	test +ve (98%)	59
		test −ve (2%)	1
ASD absent	940	test +ve (4.4%)	41
		test −ve (95.6%)	899
			1000

Looking at those who test negative, 1 will have ASD, and 899 will not, so for those with negative test results p = 1/900 = .001. Looking at those who test positive, 59 will have the disorder and 41 will not, so for those with positive test results, p = 59/100 = .59. If a child were to receive a negative test result, p would be reduced from .06 to .001 by the combination of initial knowledge with test results. If the test came back positive, p would rise from .06 to .59, suggesting that a clinical examination should be done.

We have not used Bayesian calculations in arriving at these results, but we have used prior knowledge of the frequency of ASD in the population, and of the performance of the test.

Technically, in Bayesian analysis, our estimates of how likely specific outcomes are before we gather fresh data are called our prior probabilities. If we have more than one, they make up a prior probability distribution. For short, this is often called our prior. Estimates of how likely outcomes appear after we have our data are called our posterior probabilities. If we have more than one, they make up a posterior probability distribution, often called a posterior for short.

Before seeing how we can combine a prior with data to get posterior probabilities, we need a closer look at "probability."

Frequentist and Personal Probabilities

Theorists have distinguished several types of probability, but, for our purposes we need only two: frequentist and personal probabilities. Under the frequentist conception, a probability is just the long-run relative frequency of some event. For example, life insurance companies have tables showing the probability of death at any given age for males and females, by whether they smoke. Because these tables are based on large numbers of people, gathered over a long period of time, they provide probabilities in the frequentist sense. Frequentists can also allow for estimates of long-run results worked out on theoretical grounds. For example, if we have an unbiased die, rolled in an unbiased way, landing on a surface favouring no outcome over another, the long-run probability of a "6" is just 1/6. Since we know what "unbiased" means, we can work out what our long-run results will be without having to roll the die.

Note what a strict frequentist approach rules out. We cannot speak of the "probability" of a single event, for example, the probability that Party X will win the upcoming election, or the probability that an accused committed a specific crime. We cannot because there are no long-run frequencies from which a probability can be estimated. In the same way, we are barred from estimating the probability that a theory is correct based on evidence from a few studies and a consideration of arguments pro and con. There just are no long-run frequencies in this situation.

"Personal" probabilities refer to the beliefs held by an individual. A person may well be willing to give odds for a single event, or estimate the probability that a proposition is true, based on arguments pro and con rather than on long-run frequencies. Similarly, judges routinely decide civil cases based on "the balance of probabilities." There is also a field in social science, decision theory, which is based on personal probabilities. The theory takes varying forms, but basically says that when we make decisions we estimate (however accurately) how likely various outcomes will be if we make any of the choices available to us. Given the probability of each outcome, and an estimate of its desirability, we make the choice which appears most promising.

Of course, personal probabilities can be based on long-run results; given a life insurance company's tables, and a lack of other evidence, an individual might well conclude that the probability that someone will die within a year is best estimated from that source. In Bayesian thinking, frequentist probabilities make up a subset of personal probabilities. What can be assigned probability within a frequentist framework can be dealt with inside a personal framework, but the reverse may not be true.

A potential problem is that a prior might be based on inadequate consideration. If so, after combining it with evidence the posterior might also be unrealistic. Bayesians typically respond that this is rarely an important issue in practice. First, data from a moderately large sample will "swamp" any reasonable prior. That is, quite a range of priors will lead to quite similar posterior results if there is a substantial body of data. We will see illustrations later. Second, if people report their priors, others who would have assessed the initial situation differently can redo the analysis. At least one medical journal now allows readers to go online, state their own priors, combine them with data, and come up with their own posteriors. Third, serious data analysts tend to use "vague" priors that assume no more than modest knowledge. Not wishing to be accused of using "arrogant" priors, they tend to be cautious.

A practical problem is that of defining a prior in the absence of strong previous work. A common strategy is to go with a standard prior that assumes that we know very little. There are other times when programming to do an analysis with the prior one has in mind, or a reasonable approximation, is unavailable. On the other hand, there are other situations in which the only available approach to a problem, or the only available software, is Bayesian. Certainly the range of problems that can be dealt with in a Bayesian framework has expanded rapidly in the past two decades.

Bayes' Theorem

When we have a prior, and programming is available, we combine our prior with data through Bayes' theorem. This is named after an English clergyman, Thomas Bayes, who presented a version of it in a posthumous paper (which had received some editing), and after whom the overall "Bayesian" approach is named. There are different formulae for discrete and for continuous variables. The notation for discrete variables varies. Here we will denote:

$P(A)$ = probability of having characteristic A

$P(D)$ = probability of a result in the data

$P(A / D)$ = probability of having characteristic A, given a result in the data

$P(D / A)$ = probability of a result in the data, given characteristic A

$P(AD)$ = probability of having characteristic A, and of getting a result in the data

$P(DA)$ = probability of getting a result in the data, and also having characteristic A

$P(AD)$ and $P(DA)$ refer to the same combination of events.

From basic axioms of probability,

$$P(AD) = P(A / D)P(D)$$

That is to say, the probability of having characteristic A AND there being a result in the data is the product of

the probability of A, given D

and

the probability of D.

In the example of the screening test, if the characteristic of interest (A) is having an ASD and the result in the data (D) is having a positive test score, then to solve the equation we need

$$P(A / D),$$

which we can obtain. There are 59 cases with the illness and a positive test score, and 1 without the illness and a positive score, for a total of 60 cases with a positive score. So

$$P(A / D) = 59 / 60$$

We also need P(D). We have 60 positive tests out of 1000, so

$$P(D) = 60 / 1,000$$

It also follows from basic axioms that

$$P(DA) = P(D / A)P(A)$$

This says that P(DA) is the product of

the probability of a result in the data, given the characteristic

AND

the probability of the characteristic.

We have two equations:

$$P(AD) = P(A / D)P(D), \text{ and}$$

$$P(DA) = P(D / A)P(A)$$

Their left-hand sides, P(AD) and P(DA) are equal, so their right hand sides must also be equal. Thus

$$P(A / D)P(D) = P(D / A)P(A),$$

so

$$P(A / D) = P(D / A)P(A) / P(D)$$

Rearranging,

$$P(A / D) = P(A) \frac{P(D / A)}{P(D)}$$

This is the fundamental expression of Bayes' Theorem. It applies when we are interested in a single outcome. If there are more, the denominator becomes more complicated. If the variable involved is continuous, the formula includes integrals. Here we shall focus on the most straightforward case.

On the left is the probability of having characteristic A, given the data. This probability provides the answer to the question Bayesians want to answer: given the data, what can we say about the situation in which we are interested?

On the right, we begin with P(A). This is just the probability of A before we have any data, our prior probability. It must be estimated, from previous evidence and/or careful thought. Then we have the ratio of P(D / A) to P(D). This is the ratio by which we multiply P(A) to get P(A / D).

Since the ratio gives us the factor by which we multiply P(A) to get P(A / D), we call it *Bayes' factor*.

Let us see how the formula can be used with the example data from above. The characteristic we are interested in, A, is having the illness. The result we have in our data (D) is a positive or negative test score. So, for a positive test score, P(A / D) represents the probability of having the illness, given a positive test score. On the r.h.s. of the formula, P(A) is the probability of having the illness before we know the test result, which is just .06. This is our prior probability. P(D / A) is the probability of getting a positive test score, given that you have the illness, and this we know to be .98. We need to divide P(D / A) by P(D), the probability of a positive test score, which, from the table above, is .10. Our Bayes factor, then is .98 / .10, which is far greater than one, so our posterior probability will be much greater than our prior. Substituting in

$$P(A / D) = P(A) \frac{P(D / A)}{P(D)} \text{ , we obtain}$$

$$P(A / D) = (.06) \frac{(.98)}{(.10)} = .59$$

This is just the result we obtained above, this time using a specifically Bayesian formula.

For a second example, suppose we want to estimate the probability that someone who completed first year successfully has a university degree. We are given that, for the person's age, the proportion in the population with degrees is .20. We are also given that, of those in the same age group, the proportion who reached university and completed the first year in good standing is .20. We also know that, of those with degrees, the proportion who were in good standing at the end of their first year was .80. P(A), the probability of holding a degree before we know anything about the person, is .20. P(D / A), the probability of finishing the first year in good standing for those who have degrees, is .80. P(D), the proportion who have reached the end of first year in good standing, is .20. Substituting, we have

$$P(A / D) = P(A) \frac{P(D / A)}{P(D)} \text{ ,}$$

$$= (.20) \frac{(.80)}{(.20)} = .80$$

The probability that someone who was in good standing at the end of first year has a degree is .80.

Estimating a Proportion

We should also look at an example with a continuous prior, and one which cannot be based so firmly on existing data. Let us suppose that we want to estimate the proportion of local voters who would go Conservative if an election were to be held immediately. There is no reason to suggest gaps in the probabilities, so that, for example, the figure might be .31 or .33, but no chance it might lie between those two figures. That being so, we need a continuous prior. Too, we do not wish to be accused of arrogance, of thinking we know certain results are impossible when we cannot. Thus we need a prior which assigns at least some probability to each possibility between .00 and 1.00. If we had no idea where the result might lie, we could choose a flat prior, one suggesting that, as far as we knew, one result was as likely as another.

Suppose that, having looked at recent polls and election results, we concluded it was improbable that the proportion would lie below .15, or above .55, and that our best guess was .35. We might then say we felt it was 95% probable that the true figure would lie between .15 and .50, with the mean of our prior distribution at .35, and probabilities falling off in both directions.

Figure 8.2: Prior Distributions with Means of .20, .35 and .50, and SDs of .10

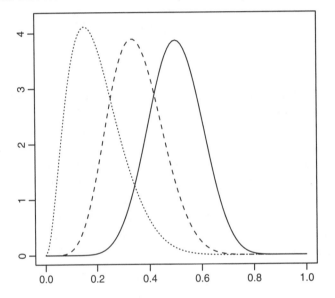

To see how much difference our choice of prior made, we could also try others.

We might consider a second set up as though it was 95% probable the true figure fell between .30 and .70, with the mean at .50. We could set up a third with its mean at .20, and the same standard deviation (.10) as the others. The three could be diagramed as shown in Figure 8.2.[1]

Given a prior, we need to combine it with data. Suppose that we obtain a sample in which 1,000 express a party preference, and 33% choose the Conservatives. When we combine this result with the priors, we obtain the posteriors shown in Figure 8.3. The data have indeed swamped the priors, so that the posterior distributions are difficult to distinguish. The posterior means corresponding to the three priors graphed are, respectively, .328, .330 and .334. A flat prior would lead to a posterior mean of .330. The difference between the highest and lowest is .006.

Figure 8.3: Posterior Distributions for Three Priors, n = 1,000

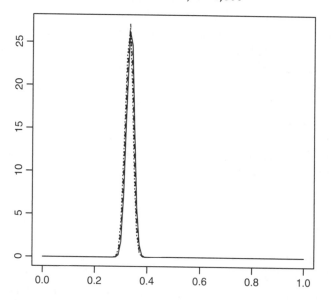

Effects of Priors and Sample Data

We could have used a considerably smaller sample and still have found the data to be much more important than the prior. With a sample of 100, differences in the posteriors are visible, as shown in Figure 8.4, but the data are clearly more important.

Figure 8.4: Posterior Distributions for Three Priors, n = 100

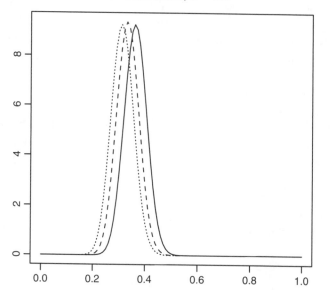

Recall that the prior means differed by as much as .50 − .20 = .30. The posteriors are, respectively, .313, .334, and .363. The largest difference is .05.

A "stronger" prior, expressing greater certainty about the electorate, would have greater influence, how much depending on the sample size. Suppose we thought it 95% probable that the true figure would fall within a range half as wide as was implied by the priors we have considered. Suppose we thought the true figure was very likely to lie between .20 and .40, with the mean of our prior at .30, and that we took a sample of 80, obtaining a mean of .23. Then the posterior mean would be .289, closer to that of the prior distribution than to that of the data. The moral is that, with large samples, the choice between reasonable (or modest) priors is not likely to matter much, but for small samples a strong prior can make a quite meaningful difference, for good or for ill, so priors require careful thought.

Credible Intervals

If we take the central 95% of the posterior, we obtain a "credible interval," a range within which we believe it is 95% probable that the true figure lies. Note the difference between the Bayesian credible interval and the standard confidence interval. The latter gives us a range within which 95% of samples will lie, if the true population figure is such-and-such. The Bayesian interval tells us it is 95% probable, given the prior and the data, that the population figure lies in a specific range.

A Bayesian analogue to a significance test can be obtained from the credible interval. If 95% of a posterior lies above zero, for example, we can say that we are 95% certain that the true figure is not zero or anything less. Suppose that a month before we took the sample of 1,000 a similar sample had been taken, but this time 30% rather than 33% had backed the Conservatives. We could easily ask whether there had been a real change or whether we were dealing with sampling fluctuation. To keep things simple, let us suppose that on both occasions we had employed a prior with mean of .35 and SD of .10 (one of the same priors used above).

We would obtain posterior means of 30.107 and 33.043, for a difference of 2.936. The credible interval is shown in Figure 8.5.

In one respect, the graph resembles those for null hypothesis testing: we see a normal curve, with shaded areas in the tails. In another respect it differs: rather than being centred around a hypothesized value (typically of zero), it is centred around 2.936, the observed difference in means for the posteriors. This shift in the graph reflects the difference between significance testing and Bayesian inference. Rather than asking how often we would observe a difference of a given size by chance, we take the data as given and ask, given the data, what difference there might have been in the world on the two occasions.

In this case, the 95% credible interval clearly includes 0, so we cannot, by this standard, rule out the possibility that there was no change during the time between surveys.

Figure 8.5: Credible Interval for Change

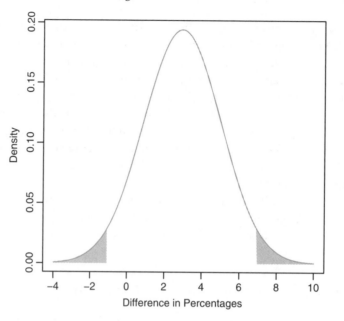

Numerical Equivalence to Standard Results

In some situations Bayesian best estimates and credible intervals are numerically identical to their counterparts under the standard approach. This is often true when we use a rectangular distribution as our prior. Doing so suggests that, across a broad range of possibilities, we really have no clear reason to expect one as compared to another. For example, if we want to estimate the mean of some variable, and we have no clear idea of what it might be, then we can set up a prior that is uniform from 0 to ± ∞. If we do, the mean of our posterior will just be the sample mean, and the SD of our posterior will be the same as the standard error of the standard approach. The difference will lie in the interpretation of the standard error. In the usual approach we would say that the standard error gives us a measure of the degree to which, if the sample mean were the true population mean, samples would vary around it. In the Bayesian approach, it gives us an indication of the range in which the true figure lies.

Since Bayesian software is now widely available, much less cautious priors can fairly readily be used. However, to go very far in this direction requires a knowledge of statistical distributions beyond the scope of this text, so we will content ourselves with an understanding of the essential logic of the Bayesian approach.

Summary

Unlike the standard model of inference, the Bayesian approach begins from what we know (or at least believe), then modifies our initial position in light of evidence. Our picture of the world before we have data is put into a prior set of probabilities (or a prior for short). Given the data, we obtain a revised picture (a posterior for short). Taking this two-step approach, we can estimate the likely difference between groups, or the likely association between variables, and we can say how far from our best estimates the truth is likely to lie.

The Bayesian approach rests on personal probabilities, beliefs held by an analyst. The analyst assesses relevant prior information to defining the personal probabilities that make up a prior. In practice, priors are often quite modest, making no strong claims about pre-existing knowledge, but this is not always so.

Given a prior, it can be combined with data through Bayes' theorem to obtain the posterior. Given a large sample, the data are likely to swamp the prior. From the posterior we can produce a credible interval, which indicates the ranges in which the true figure is likely to lie.

Review Questions on the Bayesian Alternative

1. What fundamental question does Bayesian inference try to answer that differs from the questions posed in standard inference?

2. Write the equation for Bayes' theorem used above. What do the various elements in the formula stand for?

3. What is the Bayes factor? How does it help us to estimate $P(A / D)$?

4. What is the difference between a frequentist and a personal probability?

5. What is the difference between a standard confidence interval and a Bayesian credible interval?

6. What are three replies those who accept personal probabilities have given to the argument that allowing them involves the risk that unrealistic conclusions will be drawn because of unrealistic priors?

Note

1. The beta distribution was used in setting up the priors, as it often is with proportions, because it can mimic many plausible priors.

Part III
MEASURES OF ASSOCIATION

Association and Independence

In statistics, knowing where a case stands on one variable may help us to predict where it stands on a second. If it does, we say the two are associated. If not, we say they are independent.

Suppose we have a table with these frequencies in the cells:

10	0
0	10

Here we have perfect association. Knowing which column a case is in, we can predict without error which row it is in, and vice versa.

At the other extreme, suppose we have these frequencies:

10	10
10	10

For this table, knowing the column a case is in does not help to predict which row it is in. Nor does knowing the row help to predict the column. The variables are independent.

Real data rarely show either perfect association or perfect independence. Table III.1 shows the percentage of married and unmarried respondents who reported being victims of violent crime in the preceding year (in the 2009 Statistics Canada General Social Survey.) We would not expect to predict victimization well from marital status, but a difference appears in the table.

Table III.1: Victimization by Marital Status, in Percentages		
	Married	**Others**
Victimized	20.9	26.9
Not victimized	79.1	73.1
Total	100.0	100.0

The percentages differ, so victimization and marriage do not appear to be independent, but they are weakly linked. To describe the connection between the two, we need a measure of its strength.

The measures of association we will look at take the value of 0 when there is no association, and the value of ± 1.00 when it is perfect. Degrees of association are expressed by values rising from near 0, when the variables are almost independent, to near ± 1.00, when they are almost perfectly associated.

We shall emphasize what are known as proportional reduction in error (PRE) measures. Different measures are designed for nominal, ordinal, and interval or ratio data. We shall consider measures for nominal and ordinal data in Chapter 9, and move to Pearson's r, the most widely used measure of association for interval and ratio variables, in Chapter 10.

Measures for Nominal and Ordinal Variables

Learning Objectives

In this chapter, you will learn

- how PRE measures are constructed and interpreted;

- a series of PREs designed for nominal and ordinal measures;

- how these measures are calculated;

- how an older measure, Q, is related to the PRE measure gamma, and to the odds ratio; and

- how we can use Q (like the odds ratio) to compare tables with differing marginal.

PRE Measures

In the 1960s and 1970s, previously used measures of association for nominal and ordinal variables were largely replaced in sociology and political science by Proportional Reduction in Error (PRE) measures. These included lambda, gamma, Q (a special case of gamma), and Somers' d. We shall see later that other measures are also of PRE form.

PRE measures tell us how much we can reduce our errors in predicting an outcome if we know how two variables are linked. Logically, then, for a PRE measure we must identify the type of error (or function of error) to be used. Next, we have to specify how to assess it when we don't know how the dependent variable (DV) of interest is linked to an independent variable (IV), and how to assess it when we do. Then we can calculate a proportional reduction in error measure from a formula of the form

$$PRE = (Error\ 1 - Error\ 2)\ /\ Error\ 1,$$

where Error 1 is our measure of error when we do not know how the two variables are linked, and Error 2 is our measure when we do. We divide the difference (the Reduction in Error) by Error 1 to express the reduction as a proportion of our initial error.

The measures of error we use are heavily affected by levels of measurement. We turn first to an approach for nominal data.

Lambda

Among PRE measures for nominal variables, Guttman's lambda (λ) is the most widely taught. To see how it works, we shall consider Table 9.1, based on the Canadian Election Survey of 2008. We shall consider only respondents who reported voting for one of the five largest parties.

	Region					
Party	**Atlantic**	**Quebec**	**Ontario**	**Prairies**	**BC**	**Total**
Bloc Québécois	0	261	0	0	0	261
Conservative	110	134	394	250	213	1,101
Green	34	20	60	35	38	187
Liberal	133	151	305	73	77	739
NDP	82	78	153	78	139	530
Total	359	644	912	436	467	2,818

Table 9.1: Party Voted For by Region, 2008

In this example, we will be correct if we guess someone's vote accurately, and in error if we don't. When we don't know how region and party preference are linked, all we know about party preference is found in the marginal distribution, shown on the right under the heading "Total." In the rightmost column we see that the party in the lead, the Conservatives, received 1,101 choices. Knowing nothing except the marginal distribution, we will make fewer errors if we guess that everyone chose the Conservatives than under any other strategy, so that is what we will do. We will be in error whenever someone is in another category. The number of errors equals the total (2,818) minus the number of correct guesses (1,101):

$$2,818 - 1,101 = 1,717$$

In the usual formula for lambda, this method of obtaining Error 1 is expressed as

$$(N - M),$$

[1]

where N is the number of cases in the sample, and M is the number of cases in the modal category.

Once we know how the two variables are linked, we can pick out the strongest party for each region. For the Atlantic region it is Liberal, for Quebec it is the Bloc, and elsewhere it is Conservative. Once we know this, for a given column our best strategy is to guess that everyone holds the most common preference. Table 9.2 shows how it works out.

Table 9.2: Votes for the Strongest Party, by Region		
Region	**Strongest Party**	**Votes for Strongest Party**
Atlantic	Liberal	133
Quebec	Bloc	261
Ontario	Conservative	394
Prairies	Conservative	250
BC	Conservative	213
		1,251

Our correct guesses are given by the entries in the third column, summing to 1,251. Since we have 2,818 cases, our errors come to

$$2{,}818 - 1{,}251 = 1{,}567$$

In the numerator of a PRE measure, we have Error 1 − Error 2. However, our guess for a nominal variable is either right or wrong, so the reduction in errors equals the increase in correct guesses. The increase is given by the difference between the number of correct guesses we make when we know how the variables are linked and the number we make when we don't. We have just seen that when we do, we make 1,251 correct guesses, and that when we do not we are correct 1,101 times. When we subtract we see an improvement of

$$1{,}251 - 1{,}101 = 150$$

What we have done to get the numerator of lambda can be represented by the formula

$$\Sigma\, m_j - M \,,$$

where m_j is the entry in the modal category for the jth column.

Taking the denominator from [1], a full formula, the one most typically used for lambda, is

$$\lambda = (\Sigma\, m_j - M) \,/\, (N - M)$$

Here, the numerator,

$$\Sigma\, m_j - M = 150 \,,$$

and the denominator,

$$(N - M) = 1717$$

Thus lambda = 150 / 1717 = .087.

We can interpret lambda straightforwardly.

Lambda tells us the proportion by which we can reduce our errors in guessing a case's score on the dependent variable if we know how the dependent is linked to the independent.

In the example, we have reduced our errors in predicting party preference by a proportion of .087. We typically talk about the reduction in percentage terms, though. Shifting the decimal point, we conclude that our errors have been reduced by 8.7% from their original level.

PRE measures have a well-defined meaning at all values between 0 and 1. Lambda also offers an advantage if we are able to identify one variable as independent and the other as dependent. Lambda is an "asymmetric" measure—that is, one that changes values depending on which variable is seen as affecting which. That means that we can interpret it as giving us the effect of an independent variable on a dependent. In the example, we are likely to think that regional culture and political traditions affect party preference much more than party preference affects the region people live in, so we are apt to treat region as an independent variable, and to say that it accounts for 8.7% of the variability in party preference.

Another example, based on the All Alberta survey of 1988, is found in Table 9.3.

Table 9.3: Federal Party Preference by Religious Preference

| | Religious Preference | | | | |
	Catholic	Mainline Protestant	Other Christian	Other/None	All Cases
Party Preference					
Conservative	56	105	97	41	299
Liberal	70	55	29	39	193
NDP	41	50	28	47	166
Other	27	24	52	36	139
Total	194	234	206	163	797

Calculation of Lambda.

$$\lambda = [\Sigma\, m_j - M] / (N - M)$$

$$\Sigma\, m_j - M = (70 + 105 + 97 + 47) - 299$$

$$= 319 - 299$$

$$= 20$$

$$(N - M) = (797 - 299)$$

$$= 498$$

$$\lambda = 20 / 498 = .040$$

Gamma

Gamma (γ) is designed for ordinal variables, taking advantage of our ability to say that one case scores above or below another. Unlike lambda, it is not based on how well we can predict a dependent variable; rather, we try to predict whether pairs of cases suggest a positive or a negative relationship between the variables.

In a positive relationship, as scores on one variable rise, so do scores on the other. In a negative relationship, as scores on one variable rise, scores on the other decline. If we select two cases, we may find that as scores on one rise, so do scores on the other; such a pair gives us evidence that the relationship between variables is positive. On the other hand, when scores on one rise and scores on the other fall, the pair provides evidence that the relationship is negative. We refer to pairs giving evidence of a positive relationship as concordant, and to those giving evidence of a negative relationship as discordant.

With a small sample, we can pick concordant and discordant pairs off a graph, as in Figure 9.1, which shows grades in a course by the students' GPAs coming in, at a school where F− = 0 and A+ = 13.

Figure 9.1: An Illustration of Concordant and Discordant Pairs

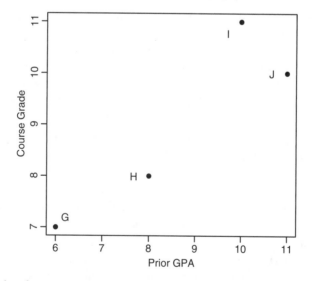

In this graph, when we pair cases G and H, as we move from one to the other, both scores rise, so this pair gives evidence of a positive relationship. The same is true for most of the other pairs: G with I, G with J, H with I, and H with J. Altogether, we have five concordant pairs. However, if we pair I with J, we see that as GPA rises, course grade falls; so this pair offers evidence of a negative relationship, and is called discordant.

With a larger sample, pair-by-pair analysis becomes unworkable. In practice, the data typically appear in tables. To see how we work with them, let us consider some data from a sample of parents whose children were in Grade 6. Each variable—parental education and parents' prediction of their child's future education—is a dichotomy, but one category is "higher" than the other, so the variables can be treated as ordinal. Their link is shown in Table 9.4.

	Table 9.4: Prediction of Child's Education by Parental Education		
	Parental Education		
Prediction of Child's Education	**No Degree**	**Degree**	**All Cases**
No Degree	246	24	270
Degree	256	58	314
	502	82	584

First, we need to identify concordant and discordant pairs. Let us take a case from cell (1,1) and another from cell (2,2). As we move from the first to the second, each variable rises, suggesting a positive relationship, and so this pair is concordant. The same will be true of any other pair involving a case from the upper left and another from the lower right.

Suppose now we pick a case from cell (2,1) and another from cell (1,2). As we move from the first to the second, parental education rises, but predictions fall, providing evidence of a negative relationship, and so this pair is discordant. The same will hold for any other pair of cases from these two cells.

Some pairs provide no evidence of a positive or of a negative relationship. If we pair a case with another from the same cell, there is no change on either variable when we go from one case to the other. If we pair a case in row 1 with a case in row 2, but in the same column, then the column variable (parents' education) shows no change, so we cannot say whether the pair is concordant or discordant. If we take a case from column 1 and a case from column 2, but in the same row, we have no change on the row variable (child's predicted education), and hence no way to declare the pair concordant or discordant.

In calculating gamma, we use only pairs that provide evidence for a positive or negative relationship. Here, any combination of cases from cells (1,1) and (2,2) yields a concordant pair, and any combination from (2,1) and (1,2) yields a discordant pair. Other cell combinations can be ignored.

Since any combination of a case from cell (1,1) with a case from cell (2,2) gives us a concordant pair, we can get the total number by multiplying cell totals. Let C be the number of concordant pairs. Then,

$$C = 246*58 = 14,268$$

Any combination of a case from cell (1,2) with another from (2,1) will yield a discordant pair, so again we can get the total number by multiplication. Let D be the number of discordant pairs:

$$D = 24*256 = 6,144$$

Getting C and D for larger tables is more complicated, and is the topic of a further example below.

Having C and D, we need to be clear how many errors we would make in predicting the direction of the same pairs if we didn't know how the two variables were linked. Without knowing about the linkage, all we could do would be to guess C and D randomly. In the long run, we would be right half the time; so for gamma,

$$\text{Error 1} = [C + D] / 2$$

Once we know how the variables are linked, we can see whether C or D is greatest. We will make fewer errors if we guess pairs are in the larger category. Here, since we have more concordant pairs, we shall guess that each pair is concordant, and we shall be in error when they are discordant. For this table,

$$\text{Error 2} = D$$

Now we can set up the formula for gamma, and calculate its value for our table. To get a PRE measure, we need

$$\frac{\text{Error 1} - \text{Error 2}}{\text{Error 1}}$$

Using $[C + D] / 2$ as Error 1 and D for Error 2, we obtain

$$Y = \frac{[C+D]/2 - D}{[C+D]/2} = \frac{[C+D] - 2D}{[C+D]} = \frac{C - D}{C + D}$$

Substituting the values we have for C and D,

$$Y = \frac{14,268 - 6,144}{14,268 + 6,144} = \frac{8,124}{20,412} = .398$$

If we have more discordant pairs than concordant, then when we know how the variables are linked we will guess that all pairs are discordant, and Error 2 will equal C. It would be possible to write

$$Y = \frac{[C+D]/2 - C}{[C+D]/2} = \frac{[C+D] - 2C}{[C+D]} = \frac{D - C}{C + D}$$

That is, we could create a second formula, with the numerator reversed. Then gamma would always be positive, because the first entry in the numerator would always exceed the second.

In fact, we always use

$$\frac{(C - D)}{(C + D)}$$

If C > D, gamma is positive. If D > C, it is negative. Using the same formula consistently means that the sign of gamma shows us whether the evidence favours a positive or a negative relationship.

Because gamma is PRE, one interpretation is that

gamma tells us the proportion by which we can reduce our errors in predicting the direction of pairs if we know how the two variables are linked.

We can convert "proportion" to percentage if we wish. In our example, a value of .398 tells us that we can reduce our errors in predicting the direction of pairs by 39.8%. If the sign of gamma had been negative, the same point would have held.

Another interpretation flows from the nature of the numerator, which gives us the excess of concordant over discordant pairs (when it is positive) or the excess of discordant pairs over concordant (when it is negative). Thus we can say that

gamma gives us the excess of one type of pair over the other, expressed as a proportion of all relevant pairs.

Of course, we can again convert "proportion" to "percentage" if we wish. Our value of .398 tells us that the excess of concordant over discordant pairs is equal to 39.8% of all the pairs considered. If gamma had been negative, we could have referred to an excess of discordant pairs over concordant.

The Sampling Distribution for Gamma

The sampling distribution for lambda, gamma, and d (to be considered next) is normal for sufficiently large samples, but the necessary N depends greatly on the strength of association. For example, when the population gamma = .00 we get a reasonable approximation to the normal with N = 10, but we need progressively large samples as the population gamma rises. The same applies to d.[1]

The point is illustrated for gamma in Figure 9.2. When the population gamma is .60, samples of 10 yield a plainly skewed distribution, but samples of 100 provide only a slight skew. Even with a population gamma of .90, a sample of 400 provides an almost normal distribution.

Figure 9.2: Some Sampling Distributions for Gamma

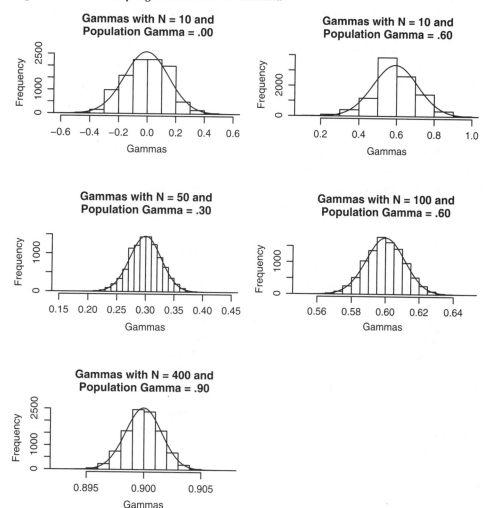

Somers' d

A modification of gamma, Somers' d, has a slightly more complicated formula:

$$d = \frac{C - D}{C + D + T_y}$$

The change lies in the addition of T_y to the denominator. It represents pairs of cases tied on the dependent variable, but not the independent. Somers believed that the effect of an independent variable should be assessed on all the cases it might have influenced. Any pair in which the cases differed on the independent variable was a possibility. However, pairs that differ on one variable but not the other are not used in gamma, so pairs tied on the dependent variable will be ignored even if they differ on the independent.

Because tied pairs are ignored, gamma can take on the value of 1.00 when many cases are off the main diagonal, as in this table.

$$10 \quad\quad 0$$
$$10 \quad\quad 10$$

Notice that in the second row there are 10 cases from the first column and 10 from the second. They are tied on the row variable, but not the column variable. There is a zero in cell b, so bc = 0, and there are no discordant pairs. The number of concordants is given by ad, here $10*10 = 100$, so

$$\textbf{gamma} = [(C - D) / (C + D)] = (100 - 0) / (100 + 0)$$
$$= 100 / 100 = 1.00$$

In Somers' view, a table like this was not an example of perfect association. His d cannot reach 1.00 unless all cases are on the diagonal.

To see the effect of bringing in cases tied on the dependent variable, let us return to the table for which we worked out gamma.

It is set up in typical form, with the independent variable as the column variable, and the dependent as the row variable. Thus, cases that differ on the IV are found in different columns, and cases tied on the DV are in the same rows in Table 9.5.

Table 9.5: Prediction of Child's Education by Parental Education			
	Parental Education		
Prediction of Child's Education	**No Degree**	**Degree**	**All cases**
No Degree	246	24	270
Degree	256	58	314
	502	82	584

We already have the values of C and D for this table. To obtain d, we only need T_y. Any pair of cases in the same row, but not in the same column, must be included in its calculation. From the first row, we can obtain

$$246*24 = 5,904$$

pairs tied on the row variable but not the column. From the second row we can obtain

$$256*58 = 14,848 \textbf{ more pairs}$$

$$\textbf{Summing, } T_y = 5,904 + 14,848 = 20,752$$

Taking C and D from the calculation of gamma, we can calculate

$$d = \frac{C - D}{C + D + T_y} = \frac{14{,}628 - 6{,}144}{14{,}628 - 6{,}144 + 20{,}752} = \frac{8{,}124}{41{,}164} = .197,$$

less than half of the value of .398 we obtained for gamma.

From their formulae it is clear that gamma and d will be the same only when T_y is zero, and since this is true only when all cases in a given row are in the same column, d is typically smaller, sometimes by a good deal.

This example illustrates another point—namely, that for 2 × 2 tables, d equals the difference in proportions between the columns. Let us compare the proportions in the upper row. For the first column, the proportion is 246 / 502 = .490, while for the second column it is 24 / 82 = .293. The difference is .490 − .293 = .197, which is the value we have for d.

A general interpretation for d can be defined from the formula. The numerator gives us the excess of concordant pairs over discordant pairs (or vice versa). The denominator includes concordant pairs, discordant pairs, and others distinguished on the independent variable but not the dependent. These categories, taken together, include all pairs distinguished on the independent variable. Thus we can say that

d gives us the excess of concordant pairs over discordant (or vice versa) as a proportion of all pairs distinguished on the independent variable.

A PRE interpretation is also possible. It is based on the assumption that, with more precise measurement, cases now tied on the dependent variable could be distinguished. It also requires the assumption that half of the pairs created by this finer measurement would be Cs and half would be Ds.

If these pairs were distinguished, then when we were trying to predict the direction of pairs, our errors on them would have to be included. As for gamma, when we do not know how the variables are linked, we can only predict which pairs will be concordant and which discordant by random guessing. Error 1, when we allow for the possibility of predicting cases now tied on the DV, but not the IV, becomes $[C + D + T_y] / 2$. Error 2 consists of all pairs we guess wrong once we know whether concordant or discordant pairs are dominant. Assuming concordant is dominant, this will be $D + T_y / 2$ (the latter because we have assumed half of the now undistinguishable pairs will be Cs and half Ds). On this basis, the formula for d becomes

$$d = \frac{[C + D + T_y] / 2 - [D + T_y / 2]}{[C + D + T_y] / 2},$$

which, if we multiply both top and bottom by 2, becomes

$$\frac{[C + D + T_y] - 2[D + T_y / 2]}{[C + D + T_y]}$$

$$= \frac{C + D + T_y - 2D - T_y}{[C + D + T_y]}$$

$$= \frac{C - D}{C + D + T_y}$$

This being so, we can say that

d gives us the proportion by which we could reduce our errors in predicting the direction of pairs, if we had a precise measure of the DV, if the pairs created by this precise measurement were evenly divided into concordant and discordant, and if we knew how the two variables were linked.

Since the PRE interpretation requires additional assumptions, the former has the advantage in simplicity, but you will certainly see PRE interpretations.

Since d includes pairs tied on the DV but not the IV in the denominator, its value depends on which variable is treated as the DV and which the IV. If we are clear on which variable causes which, it is good to be able to choose the value that shows the effect of the IV on the DV.

In this respect, d has an advantage over gamma, which takes on the same value whichever variable is independent. Technically, we say that gamma is symmetric, meaning that it gives the same value whichever way we view the variables. Symmetry may be advantageous when we do not know which should be treated as independent, but is a disadvantage otherwise.

Calculating Gamma and d for Tables Larger than 2 × 2

To see how gamma and d are calculated for a table larger than 2 × 2, we shall use the data in Table 9.6.

Table 9.6: Frequency of Feeling the House Is Inadequate for Guests, by Poverty Status				
	Degree of Poverty			
	Very Poor	**Poor**	**Borderline**	**Total**
Frequency				
Often	10	6	1	17
Sometimes	12	9	6	27
Rarely, never	18	23	24	65
Total	40	38	31	109

To calculate gamma, we need the number of concordant pairs. Scores on the IV increase as we move to the right, scores on the DV increase as we move down. In a table set up this way, concordant pairs are obtained from cells located so that one is below and to the right of the other. We must identify all such pairs of cells. For each, we shall have to multiply the cell frequencies to get the number of concordant pairs that can be created from them.

We shall begin in the upper left-hand corner. Cell (1,1) is above and to the left of four other cells, (2,2), (2,3), (3,2), and (3,3). Thus, we may pair (1,1) with any of these. To get the number of concordant pairs this creates, we just multiply

$$10(9 + 6 + 23 + 24) = 620$$

(The bracketed numbers are the frequencies in the other cells.)

Moving to the right, cell (1,2) is above and to the left of two other cells, (2,3) and (3,3). We can obtain concordant pairs by matching (1,2) with each of them. To obtain the number, we take

$$6(6 + 24) = 180$$

Moving to cell (1,3), we find no cells to its right. We move to the second row.

We begin again at the left, with cell (2,1). There are two cells below it and to its right, cells (3,2) and (3,3). We can get further concordant pairs by taking

$$12(23 + 24) = 564$$

Cell (2,2) can be paired with one cell, cell (3,3). The number of concordant pairs is

$$9(24) = 216$$

Cell (2,3) has no cells to its right. If we go to the third row, there are no cells below. C, the number of concordant pairs, is the sum of those we have identified:

$$C = 620 + 180 + 564 + 216 = 1580$$

We need the number of discordant pairs as well. For a table set up with scores on the IV rising as we move right and scores on the DV rising as we descend, we can get discordant pairs by matching cells so one is below and to the left of the other. We begin on the upper right, with cell (1,3). There are four cells below it and to its left, (2,1), (2,2), (3,1) and (3,2). We get the number of discordants created from these cells from

$$1(12 + 9 + 18 + 23) = 62$$

Moving to the left, cell (1,2) can be matched with two cells below it and to its left, (2,1) and (3,1). To get the number of discordant pairs, we take

$$6(12 + 18) = 180$$

Cell (1,1) has no cells to its left, so we move to the second row. On the right, cell (2,3) can be matched with two cells below and to the left, (3,1) and (3,2). To get the discordants, we multiply

$$6(18 + 23) = 246$$

Moving to the left, cell (2,2) can be matched with (3,1). We multiply

$$9(18) = 162$$

Cell (2,1) has no cells to its left, and cells in the third row have none below, so

$$D = 62 + 180 + 246 + 162 = 650$$

The numerator for gamma is (C − D). We have C = 1,580, and D = 650, so the numerator is 1580 − 650 = 930. The denominator is (C + D), or 1580 + 650, which comes to 2,230, so

$$\textbf{gamma} = \textbf{930 / 2,230} = \textbf{.417}$$

For d, we need the number of pairs tied on the DV, but not on the IV. To be tied on the DV, cases must be in the same row. To be in different categories of the IV, they must be in different columns. So we match cells which are in the same row. In the first row, we can pair (1,1) with (1,2) and with (1,3). We get the number of pairs tied on the DV and not the IV by taking

$$10(6 + 1) = 70$$

In row one, we can also match (1,2) with (1,3) obtaining

$$6(1) = 6 \textbf{ pairs}$$

In the second row, we begin similarly, matching (2,1) with (2,2) and (2,3), obtaining

$$12(9 + 6) = 180 \textbf{ pairs}$$

Then we match (2,2) with (2,3) to get

$$9(6) = 54 \textbf{ pairs}$$

We proceed the same way with the third row, matching cell (3,1) with (3,2) and (3,3), to get

$$18(23 + 24) = 846 \textbf{ pairs}$$

Then we match (3,2) with (3,3) to get

$$23(24) = 552 \textbf{ pairs}$$

Having identified all the pairs matched on the DV but not the IV, we take

$$T_y = 70 + 6 + 180 + 54 + 846 + 552 = 1708$$

We have already obtained (C − D) = 930 and (C + D) = 2,230. We can now calculate

$$\textbf{d} = \textbf{(930) / (2,230 + 1,708)} = \textbf{.236}$$

Yule's Q as a Special Case of Gamma

Although invented half a century earlier, Q, a measure due to the pioneering British statistician Yule, is a special case of gamma. In fact, the example we looked at could equally be seen as an example of the use of Q, which is equal to gamma for a 2 × 2 table.

Although, for a 2 × 2 table, Q has the same value as gamma, the formula is different because the idea of concordant and discordant pairs had not been developed at Yule's time.

$$Q = (ad - bc) / (ad + bc),$$

where a is the value in cell (1,1), b the value in (1,2), c the value in (2,1) and d the value in (2,2). Using this code for the cells, a 2 × 2 table looks like this:

a	b
c	d

To illustrate the equivalence of Q and gamma (for 2 × 2 tables), we can return to the table for which we worked out gamma, now presented in Table 9.7.

Table 9.7: Prediction of Child's Education, by Parental Education

	Parental Education		
Prediction of Child's Education	**No Degree**	**Degree**	**All Cases**
No Degree	246	24	270
Degree	256	58	314
	502	82	584

Here a = 246; b = 24; c = 256; and d = 58, so

$$Q = (ad - bc) / (ad + bc)$$
$$= [246(58) - 24(256)] / [246(58) + 24(256)]$$
$$= (14{,}268 - 6{,}144) / (14{,}268 + 6{,}144)$$
$$= (8{,}124) / 20{,}412 = .398$$

This is the same value obtained above. If you check back, you will see that the same numbers are involved through the full calculation. This is because C in the formula for gamma equals ad in the formula for Q, and D equals bc. This will hold for any 2 × 2 table in which column scores rise as we move right and row scores rise as we move down.

Although gamma and Q are numerically identical, a difference in usage does appear. Yule did not restrict his statistic to ordered variables, and Q has been used frequently with nominal dichotomies. We can assign direction to pairs arbitrarily, as long as everyone knows what we have done so the results are correctly interpreted. Consider Table 9.8, in which criminal victimization is predicted from marital status. (We have seen this table above, but in percentaged form.) Both variables are nominal.

Table 9.8: Victimization by Marital Status

	Marital Status		
	Married	Other	Total
Victimization Status			
Victimized	1,927	2,748	4,675
Not victimized	7,298	7,450	14,748
	9,225	10,198	19,423

Here a = 1,927; b = 2,748; c = 7,298; and d = 7,450.

$$Q = (ad - bc) / (ad + bc)$$

$$= [1,927(7,450) - 2,748(7,298)] / [1,927(7,450) + 2,748(7,298)]$$

$$= -.166$$

Because the variables are nominal, we could easily reverse the coding, and then interchange the rows or columns. To illustrate what happens, we reverse the rows for victimization in Table 9.9.

Table 9.9: Victimization by Marital Status, with Rows Reversed

	Marital Status		
	Married	Other	Total
Victimization Status			
Not victimized	7,298	7,450	14,748
Victimized	1,927	2,748	4,675
	9,225	10,198	19,423

In the first table, ad took on the value 1,927(7,450), whereas in the second bc = 7,450(1,927). That is, the value of ad from the first becomes the value of bc in the second. In the same way, the value of bc from the first (2,748 × 7,298) becomes that of ad in the second.

Substituting values from the revised table, we get

$$Q = (ad - bc) / (ad + bc)$$

$$= [7,298(2,748) - 7,450(1,927)] / [7,298(2,748) + 7,450(1,927)]$$

$$= .166$$

In the original table, a negative sign meant that being unmarried was linked to victimization. In the modified table, a positive sign means the same thing. In general, if the categories are switched, for either variable, the sign of Q reverses. As long as we understand the meaning of the sign, everything is fine. If we had three or more categories in either variable, we could not say what gamma, calculated for nominal variables, would mean, but for a 2 × 2 table, we can calculate Q (a special case) and everything works out.

Q's interpretation is either equivalent to or very similar to that of gamma. For ordinal variables, it is equivalent. For nominal variables we can interpret it without reference to ordering of the pairs.

As an example, let us use the value of .166 obtained for the second victimization table. Since it is positive, it implies that ad > bc. That is, it implies that the pairs associating victimization with being unmarried and non-victimization with being married exceed the pairs making the opposite association (of marriage with victimization and being unmarried with non-victimization). Further, it tells us that pairs linking victimization with being unmarried exceed pairs linking it with being married by 16.6% of all pairs examined. That is,

Q can be interpreted as the excess of pairs favouring one type of association over those favouring the other as a proportion of all pairs favouring one or the other.

A kind of PRE interpretation is also possible. If we knew nothing of how victimization was linked to place of residence, we could only guess whether a pair would suggest victimization was linked to urban or to rural living by a coin toss. In the long run, we would be right half the time and wrong half the time.

Q gives us the proportion by which we can reduce our errors in guessing whether pairs favour one form of association (or the other) if we know how the variables are linked,

and this is so whether there is an ordering to the pairs or not.

In our example, we can just say

the pairs linking victimization to being unmarried exceed those linking it to being unmarried by 16.6% of all such pairs; and

we can reduce our errors in predicting whether a pair links victimization to one status or the other by 16.6%.

In short, when we have nominal variables, our interpretation parallels that for ordinal variables, but need not make reference to any ordering of the categories.

As we shall see later, Q is not only useful as a measure of association for two dichotomies, but also in comparing the association of two variables in separate tables. First, though, we need to look at how Q is related to another measure, the odds ratio.

The Odds Ratio

Suppose we have a 2 × 2 table whose cell values are these:

5	5
5	5

For cases in the first column, the odds on being in the upper row, as opposed to the lower row, are 5 to 5. The odds on being in the upper row are the same for cases in the second column, 5 to 5. Suppose we want to compare the odds on being in the upper row for cases in column 1 and cases in column 2. We can express the odds for the first column as 5 / 5. We can express the odds for the second column the same way, as 5 / 5. We can compare these two odds by setting them into a ratio. We obtain the odds ratio (OR) by taking

$$(5 / 5) / (5 / 5)$$

That is, we divide the odds for the first column by the odds for the second. In this case the two are the same, so we finish with an odds ratio of one.

When the two variables are independent, the odds on being in row 1, as opposed to row 2, are the same for both columns. Thus the odds ratio is one. It goes up or down depending on the difference in the two odds. Suppose we have the table

$$\begin{array}{cc} 6 & 4 \\ 4 & 6 \end{array}$$

Here the odds on being in the first row, for column 1, are 6 / 4. The odds for column 2 are 4 / 6. We obtain the ratio as before, by taking

$$(6 / 4) / (4 / 6) = (6 / 4)(6 / 4) = (6 \times 6) / (4 \times 4) = 36 / 16 = 9 / 4 = 2.25$$

The odds on being in row 1 are 2.25 times higher if a case is in column 1 than if it is in column 2.

Suppose the columns are interchanged, giving

$$\begin{array}{cc} 4 & 6 \\ 6 & 4 \end{array}$$

Now the odds ratio is given by

$$(4 / 6) / (6 / 4) = (4 / 6)(4 / 6) = (4 \times 4) / (6 \times 6) = 16 / 36 = 4 / 9 = .444$$

When the odds on being in row 1 are lower for column 1 than for column 2, we have a ratio of less than one. If we interchange rows or columns, we get the reciprocal of the initial odds ratio. For the original table we had 9 / 4, and after interchanging columns we have 4 / 9.

Let us denote the cell values as a, b, c, and d, as shown below.

$$\begin{array}{cc} a & b \\ c & d \end{array}$$

Then the odds ratio can be written

$$\frac{a/c}{b/d} = [a/c] \times [d/b] = \frac{ad}{bc}$$

The OR is often called the cross-product ratio, because it can be obtained from the products of diagonally opposite cell values (ad and bc).

The OR can be used when we want to know how an intervention affects an outcome. Suppose we want to know whether students are less likely to be in special education if we provide a program to improve their life chances, and we have the results below in Table 9.10.

	Table 9.10: Use of Special Education by Type of Site	
	Type of Site	
	Program Site	**Comparison Site**
Outcome		
In special education	78	37
Not in special education	231	90

If we want to compare the odds for the program site to the comparison site directly, we can calculate them for each column. On the left, for the program site, we have

$$78 / 231 = .338$$

For the comparison site, we have

$$37 / 90 = .411$$

We can then put them into a ratio: .338 / .411 = .821

To get the OR without calculating the odds for each column we take

$$ad / bc = [(78)(90)] / [(37)(231)] = 7020 / 8547 = .821$$

The odds on use of special education are lower for the program site by a factor of .821.

Odds ratios are very common in the medical literature, because key variables are dichotomous (did someone become ill or not; receive treatment or not; recover without treatment or not; respond to therapy or not; or live or die). In the social sciences, we also have a good number of dichotomous dependent variables that can be combined with dichotomous predictors. For example, in Table 9.11 we have data on whether a couple with children broke up between interviews in a longitudinal study. The predictor is whether they were initially married or in a common-law union (CLU).

Table 9.11: Breakup between Interviews by Marital Status

	Marital Status		
Breakup	CLU	Married	All Cases
Yes	42	264	306
No	127	2,170	2,297
Total	169	2,434	2,603

For the first column, the OR is 42 / 127 = .331. For the second, it is 264 / 2,170 = .122. Placing them in a ratio, we obtain 2.718.

We can also get the ratio directly, from

$$(42)(2,170) / (264)(127) = 2.718$$

However we calculate the ratio, the odds on a breakup are 2.718 times higher for those in a common-law union than those who are married.

Another use for the OR is to assess the affinity between groups or categories. For example, it can be used to examine homogamy. In Table 9.12 we have data showing the tendency of those who have not already been married to marry others who have never been married in Canada in 2002.

Table 9.12: Groom's Prior Marital Status by Bride's Prior Status

	Bride	
Groom	Never Married	Other
Never married	96,924	13,068
Other	13,971	22,775

If the bride has not been married before, the bulk of grooms have not either, while if the brides are in the Other category, which includes the divorced and widowed, most of the grooms are as well. For previously unmarried brides, the odds on the groom's being previously unmarried are

$$96,924 / 13,971 = 6.938$$

For brides married before, they are

$$13,068 / 22,775 = .574$$

As above, we can get the odds ratio by dividing the odds for one column by the odds for the other, taking

$$6.938 / .574 = 12.091$$

Or we can get the OR directly, from

$$\text{ad / bc} = (96{,}924)(22{,}775) / (13{,}068)(13{,}971) = 12.091$$

The OR of 12.091 is the factor by which we multiply the odds on the groom's being previously single if the bride has never been married. Put another way, the odds on the groom's being previously single, if the bride has not been married, are more than 12 times higher than if the bride has been married.

Q as a Function of the Odds Ratio

To calculate Q we take (ad − bc) / (ad + bc), whereas for the OR we take ad / bc. Each is built up from the same elements, ad and bc, and in fact each is a straightforward function of the other.

Suppose we begin with the formula for Q,

$$Q = \frac{\text{ad} - \text{bc}}{\text{ad} + \text{bc}},$$

then divide both the numerator and the denominator by bc. We get

$$\frac{\dfrac{\text{ad}}{\text{bc}} - \dfrac{\text{bc}}{\text{bc}}}{\dfrac{\text{ad}}{\text{bc}} + \dfrac{\text{bc}}{\text{bc}}} = \frac{\dfrac{\text{ad}}{\text{bc}} - 1}{\dfrac{\text{ad}}{\text{bc}} + 1} = \frac{\text{OR} - 1}{\text{OR} + 1}$$

If we have the OR, this result provides a formula for obtaining Q. We substitute in

$$Q = (\text{OR} - 1) / (\text{OR} + 1)$$

and follow the algebra through. In the special education example above, we obtained OR = .821. So

$$Q = (\text{OR} - 1) / (\text{OR} + 1) = (.821 - 1) / (.821 + 1)$$

$$= -.179 / 1.821 = -.098$$

If we have Q, we can get the OR with just a few more steps.[2]

The Usefulness of Q with Changing Marginals

Because Q is a function of the odds ratio, Q shares one of its desirable features, its independence of marginal totals. Other measures, including percentage differences and d, do not have this property. Its importance can best be seen by illustration. Consider a table, for which the cell frequencies are

40	20
60	80

Here the odds ratio is

$$\mathbf{ad / bc} =$$

$$(40)(80) / (20)(60) =$$

$$3200 / 1200 = 32 / 12 = 8 / 3 = 2.667$$

Now consider another table, in which the numbers in the first row are halved, and the second row remains as before:

20	10
60	80

$$\mathbf{ad / bc} = [20(80)] / [10(60)] = 1,600 / 600 = 16 / 6 = 8 / 3 = 2.667$$

The OR has not changed. More generally, we can multiply the entries in either row or either column by any non-zero number and the odds ratio will not be affected.

Suppose, on the other hand, we want to compare the percentages in the two columns of the table. For the original table, because the column totals are 100s, the percentages are the same as the cell values.

40	20
68	80

For the revised table, we get

25.0	11.1
75.0	88.9

For the original table, the difference in percentages in the first row is $(40 - 20) = 20$. For the revised version the difference is

$$(25.0 - 11.1) = 13.9$$

The percentage difference is not stable under changing row marginals. As a result d, which for a 2 × 2 table equals the percentage difference, is not stable either, falling from .200 to .139.

The stability of the OR and Q under changing marginals reflects their exclusive focus on the inner structure of a table, in the sense of the ratios among the cell values within it. As we have seen, measures not based on the OR are affected not just by the inner structure, but also by the marginals. If we wish to compare 2 × 2 tables with differing marginal proportions, we should give careful consideration to using the odds ratio or Q. Using one of them, we will not have to wonder to what extent differences in a measure of association result from changing marginals.

We will take advantage of Q's focus on inner structure in our later consideration of conditional tables. For now, we will briefly examine two tables to see how Q can be used. The first, Table 9.13, shows the link between single parenthood and smoking status (smoker or non-smoker), in percentages.

	Table 9.13: Smoking Status by Parental Status, in Percentages		
	Parental Status		
Smoker	**Single Parent**	**Partnered Parent**	**All Cases**
Yes	31.4	16.3	20.1
No	68.6	81.7	79.9
Total	100.0	100.0	100.0
N	4,980	31,948	36,928

Q = .343

The row marginals change considerably when we break the sample into the native-born and immigrants. In Table 9.13, for all cases, the percentage of smokers is 20.1. In Table 9.14, for the native-born, it is 22.7, and for immigrants it is 13.1. We are fortunate to be able to work with Q, which does not depend on row proportions.

	Table 9.14: Smoking Status by Parental Status by Nativity, in Percentages							
Native-born				**Immigrants**				
Smoker	**Single Parent**	**Partnered Parent**	**All Cases**	**Smoker**	**Single Parent**	**Partnered Parent**	**All Cases**
Yes	36.3	20.5	22.7	Yes	18.3	12.8	13.1
No	63.7	79.5	77.3	No	81.7	87.2	86.9
Total	100.0	100.0	100.0	Total	100.0	100.0	100.0
N	3,608	22,336	25,944	N	1,240	8,912	10,152

Q = .337   Q = .206

For the full sample, Q = .343. Its positive sign implies that the heavy cells link smoking to single parenthood and non-smoking to partnered parenthood. Since Q is based only on the inner structure of the table, we can compare its value for the full sample to the .337 for the native-born, and the .206 for immigrants. By this comparison, the link between smoking and single parenthood is weaker for immigrants than for the native-born. Notice that the percentage of smokers among immigrants, at 13.1, is more than a third smaller than the percentage for the native-born, at 22.7. This difference in the marginal proportions would cause a problem for a comparison not based on the inner structure of the table, but we can compare values of Q straightforwardly.

Summary

In this chapter, we have turned our attention to measures of association between variables, in particular to proportional reduction in error (PRE) measures. We have looked at lambda (λ), a measure for nominal variables, and at gamma (γ) and d, two related measures for ordinal variables. We have given attention to Q, which for 2×2 tables is numerically identical to gamma. It is a function of the odds ratio (OR), another useful measure for 2×2 tables. Q thus shares a desirable property of the OR, its independence from marginal totals. Because of this, Q can be used to compare the association in tables with different marginal proportions. It will be used for this purpose in Chapter 12, on conditional tables.

Review Questions on Measures for Nominal and Ordinal Variables

1. What is a PRE measure?

2. For Table 9.15, would it make sense to use lambda as a measure of association? Why? How about gamma? d?

Table 9.15: Party Voted For in 2008 by Party Voted For in 2006, in Percentages

Vote 2008	Vote 2006					
	Bloc	**Conservative**	**Green**	**Liberal**	**NDP**	**Total**
Bloc	84.6	1.7	5.8	1.0	1.7	11.6
Conservative	5.3	80.4	7.2	15.9	7.9	37.4
Green	.6	3.4	46.4	4.8	8.8	6.2
Liberal	6.4	7.4	20.3	65.3	13.0	26.5
NDP	3.2	7.1	20.3	13.0	68.6	18.4
Total	100.1	100.0	100.0	100.0	100.0	100.1
N	188	593	69	484	239	1573

X^2(16 DF) = 2467.889, p < .001, lambda = .569, p < .001

3. For the table above, how would you obtain Error 2 for lambda?

4. For this table, lambda = .569. What does this value tell us?

5. Suppose that the political party preferences of students are predicted from their fields of study. Would it be correct to use lambda as a measure of association here? Why? What, if anything, would a lambda of .15 mean? What, if anything would d mean?

6. In statistics, what is the difference between a positive and a negative relationship?

7. In Table 9.16, how would you identify concordant pairs?

Table 9.16: Distance Run by Level of Aerobic Exercise

	Level of Aerobic Exercise	
	Gets Little Aerobic Exercise	**Gets Much Aerobic Exercise**
Distance Run		
Runs < 2k in 12 min.	30	10
Runs 2K+ in 12 min.	10	30

8. For the table above, set up the calculations to find the odds ratio. You don't have to do it: just show you know how to go about the job.

9. What is the odds ratio? What are some of its uses?

10. Set up the calculations to get Q. What is the relationship between Q and the odds ratio?

11. What is the relationship between the odds ratio and the marginal totals? Between the odds ratio and the "inner structure" of the table?

12. For the table above, what is the value of gamma? (Or Q?) What is one interpretation of its value?

13. What are two ways to interpret Q? When does it equal gamma?

14. Give the formula for d and indicate what the various symbols in it refer to.

15. What is the difference between gamma and d?

16. Why did Somers wish to suggest an alternative to gamma?

17. Are gamma and d PRE measures? If so, what sorts of error do they look at?

18. Suppose we calculated d and obtained .600. How would we interpret its value? Why would it differ from gamma?

19. What is the difference between a symmetric and an asymmetric measure of association? Give an example of each.

20. Which of the statistics considered in this chapter are PRE?

21. Why do lambda, gamma, and d not have the same numeric values?

22. gamma = .618. What are two ways to interpret this value?

23. d = .379. What are two ways to interpret this value?

Notes

1. The two measures have the same numerator (C − D), and typically statistical software works from the sampling distribution of the numerator in testing the significance of each of them.

2. $-.098 = (OR − 1) / (OR + 1)$

    We multiply by (OR + 1), obtaining

    $$-.098(OR + 1) = OR − 1$$

    $$-.098(OR) −.098 = OR − 1$$

    We move OR left and −.098 right, getting

    $$-.098(OR) − OR = −1 + .098$$

    $$-1.098(OR) = −.902$$

    We divide both sides by −1.098, so that

    $$OR = −.902 / (−1.098)$$

    $$OR = .821 \text{, the value obtained above}$$

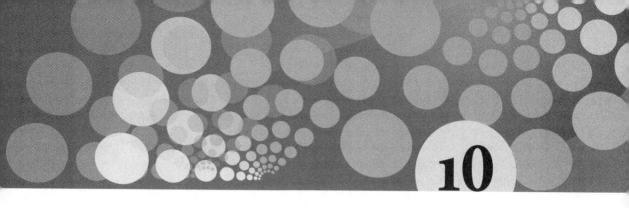

Pearson's r

Learning Objectives

In this chapter, you will learn

- the logic behind the formula for r;
- why it is a measure of linear association;
- two ways to interpret it;
- its sampling distribution;
- two special cases with their own names; and
- how we graph data to which r is applicable.

Explaining the Formula

We have examined the most widely used measures of association for nominal and ordinal variables. Pearson's r is the one most widely employed with interval and ratio variables. Many different formulae have appeared, all yielding the same value. One well suited to explaining r is

$$r = \frac{\sum (x_i - \overline{x})(y_i - \overline{y}) / N}{[\sum (x_i - \overline{x})^2 / N]^{.5} [\sum (y_i - \overline{y})^2 / N]^{.5}}$$

There is nothing new about the two quantities in the denominator; they are standard deviations, of x and y respectively.[1] The numerator is called the Covariance. It gets the name in part because it is similar in form to a Variance. Note that if $(y_i - \overline{y})$ is replaced with $(x_i - \overline{x})$, we have

$$\sum (x_i - \overline{x})(x_i - \overline{x}) / N = \sum (x_i - \overline{x})^2 / N ,$$

the Variance of x. Similarly, if $(x_i - \bar{x})$ is replaced with $(y_i - \bar{y})$, we have

$$\Sigma\,(y_i - \bar{y})\,(y_i - \bar{y})\,/\,N = \Sigma\,(y_i - \bar{y})^2\,/\,N\,,$$

the Variance of y. The other reason why the numerator is called the Covariance is that it reflects how x and y vary <u>together</u>, or covary.

In the Covariance, we sum the products

$$(x_i - \bar{x})(y_i - \bar{y})$$

If the value of x for a given case equals $\bar{x}$, then $x_i - \bar{x} = 0$, so the product $(x_i - \bar{x})(y_i - \bar{y})$ will be 0. Similarly, if $y_i - \bar{y} = 0$, the product will be 0. If each variable scores above its mean, $(x_i - \bar{x})$ will be positive and so will $(y_i - \bar{y})$. Then the product will be positive.

If each scores below its mean, $(x_i - \bar{x})$ will be negative and so will $(y_i - \bar{y})$, so the product $(x_i - \bar{x})(y_i - \bar{y})$ will be positive. Thus, observations which suggest a positive association, which are illustrated in the graph below, make a positive contribution to the Covariance.

On the other hand, if x is above its mean and y below, or vice versa, their product will be negative. Thus observations which suggest a negative association make a negative contribution to the Covariance. Observations of each kind are presented in Figure 10.1.

Figure 10.1: Points Suggesting Positive and Negative Correlation

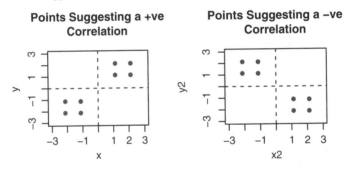

Notice also that using $(x_i - \bar{x})$ and $(y_i - \bar{y})$ gives greater weight to cases farther from the means, which provide stronger evidence for the direction of association than those close to the centre of the data.

When we sum the contributions of individual cases, three things may happen:

1. The positive may exceed the negative. Then the Covariance will be positive. This will indicate an overall positive association between the two variables.

2. The negative contributions may exceed the positive. Then the Covariance will be negative. This will indicate an overall negative association between the two variables.

3. The negative and positive contributions may cancel, so that the Covariance is zero. This will indicate that there is neither a positive nor a negative association between the two variables.

The denominator of r includes only the two standard deviations, both of which must be positive. Thus the numerator determines the sign of r, which is positive if the evidence for a positive relationship outweighs that for a negative, and negative if the evidence for a negative association outweighs that for a positive.

The SDs in the denominator ensure that r will remain within the range from −1 to +1. When r equals +1, there is a perfect positive association. When r is −1, there is a perfect negative association. If it is 0, we say there is no linear association.

We say no linear association because r simply balances out the positive and negative contributions made by individual observations. Sometimes these balance perfectly, even though there is a clear non-linear association between two variables. This happens, for example, when the observations form a circle on a graph. It can also happen when the graph rises over part of its range and declines over another. Anscombe's (1973) examples, graphed in Figure 10.2, illustrate other ways in which we may be misled if we use r for non-linear relationships. For each graph, r is .816 or .817, but r is a good summary only for the upper left-hand graph, in which we see a reasonably straightforward linear association.

Figure 10.2: Four Data Sets with Almost Identical Values of r

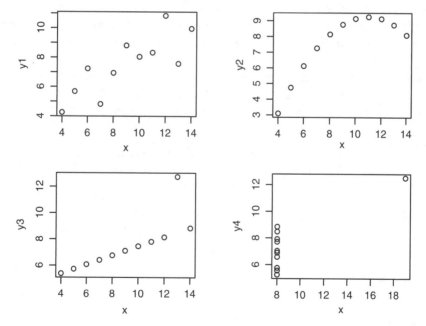

As well as in algebraic notation, the formula for r may be written

$$r = \frac{\text{Cov(x,y)}}{\text{SD(x)SD(y)}},$$

where Cov(x,y) stands for the Covariance of x and y.

Another formula reflects the situation when the variables have been standardized. Recall that when we standardize a variable, its SD becomes 1. If we standardize both x and y, both SDs become one so the formula above becomes

$$r = \frac{Cov(x,y)}{(1)(1)} = Cov(x,y)$$

Following usual practice, we can denote the standardized x and y as z_x and z_y. Since the mean of a standardized variable is 0,

$$r = Cov(z_x, z_y) = \Sigma (z_{xi} - 0)(z_{yi} - 0) / N$$
$$= \Sigma (z_{xi})(z_{yi}) / N$$

In words, r becomes the mean of the products of x and y, once these have been standardized.

Calculation

In practice, we obtain r through software, which typically uses a formula different from those we have seen, but which yields the same result. As it may help in understanding the measure to see it worked out, we will take an example. So we can concentrate on the process rather than the numbers, we shall use a fictitious set of grade averages coming into a course as x and course grades as y. Table 10.1 is a worksheet showing how we might calculate r through the formula above, $r = \Sigma(z_{xi})(z_{yi}) / N$.

We have seen that to standardize x, we take $(x_i - \bar{x}) / SD(x)$. For y, we take $(y_i - \bar{y}) / SD(y)$. In practice we use software to get z-scores, so the calculations to get them will not be shown. Fewer decimal digits than were used in the calculation will be presented.

Table 10.1: Worksheet to Illustrate the Calculation of r				
x	y	z_x	z_y	$z_x z_y$
60	65	−1.37	−.90	1.23
70	72	−.39	−.11	.04
79	80	.68	.79	.54
83	75	1.13	.23	.26
				2.07 ÷ 4 = .52

Interpretations of r

Pearson's r has been interpreted in a variety of ways, two of which are particularly useful to social scientists. One is based on the fact that squaring it yields r^2, a PRE measure. Suppose, for example, we correlate years of education and income, obtaining r = .35, and r^2 = .1225. We conclude that years of education will explain 12.25% of the Variance in income. More generally, if we square r we can interpret the result as the proportion of the Variance in one variable that can be accounted for from the other.

The second is based on what happens if we standardize the variables in bivariate regression, a topic to which we will soon turn. As it happens, r tells us how much of an SD of change we get in one variable, on average, for one SD of change in the other. So if, as above, r = .35, we can say that, on average, we get .35 SDs of change in income for one SD of change in years of education.

The Sampling Distribution of r

As sample size increases, the sampling distribution of r tends to the normal. However, it does so much more quickly when r is close to zero than when it is larger. To speed up the convergence to the normal, r can be transformed to

$$z = .5 \ [\ln((1 + r) / (1 - r))]$$

For larger samples, or for samples from populations where r is low, the transformation is not so important, as shown in Figure 10.3. With a population r of .00, an N of 30 yields a reasonable approximation to the normal. With a population r of .71, we need a much larger sample: an N of 400 yields only a very slightly skew. With the transformation, an N of 30 works fairly well for r = .71.

Figure 10.3: Sampling Distributions for r

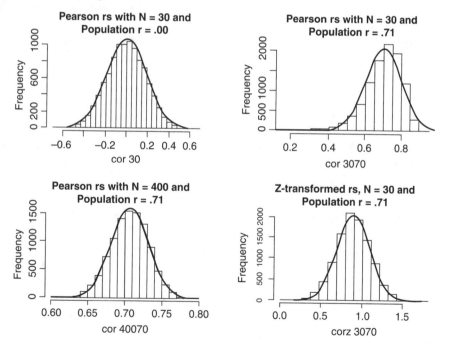

Spearman's Rho

Above we have seen two measures of association for ordinal variables, gamma and d. An earlier measure, still seen, is readily interpreted, because it is simply a variant of r. Spearman's rho, denoted by the lower case Greek ρ to distinguish it from r, is actually a special case.

We calculate ρ just as r, except that instead of using the observed values of x and y we use their rank positions (1 though N). If several scores lie in the same category, we suppose that, with finer measurement, they could be distinguished, and we take the median rank that would then be found for the set. Suppose, for example, that we have letter grades for two first-year classes and that we want to see how highly they correlate. For one class, we might have the results shown in Table 10.2.

Table 10.2: Hypothetical Class Results			
Grade	Frequency	Ranks	Median
A	13	1–13	7
B	22	14–35	24.5
C	40	36–75	55.5
D	18	76–93	84.5
F	7	94–100	97.5

We could calculate the median scores for the second class in the same way, then calculate r using the median ranks in place of the values of x and y. In practice, of course, we would be pleased to allow software to carry out the task.

The interpretation of ρ, since it is a special case of r, is the same, except that we are referring to the ability of one set of ranks to predict another, rather than the ability of one set of raw scores to predict another.

Phi

Phi (φ) is another special case. For a 2 × 2 table, $\varphi = \sqrt{X^2 / N}$. Although it is calculated from chi-square, it is equal to r, calculated in the usual way, for the same data. It appears less in some other fields of social science than it did once, but continues to make regular appearances in psychology.

Correlation Matrices

The links among several variables may be compactly displayed in a correlation matrix. When we are working with scales, correlation matrices can give us an initial sense of how our items are performing. If they are measuring the same thing, they ought to correlate positively (unless some have been worded in reverse, so a high score means what a low score would mean for other items.)

The matrix in Table 10.3 is based on six items from a scale for perceived social support, presented below.

1. If something went wrong, no one would help me. (reverse worded)

2. I have family and friends who help me feel safe, secure and happy.

3. There is someone I trust whom I could turn to for advice if I were having problems.

4. There is no one I feel comfortable talking about problems with. (reverse worded)

5. I lack a feeling of intimacy with another person. (reverse worded)

6. There are people I can count on in an emergency.

Reverse worded items have been reflected—that is, recoded so high scores are low and vice versa—so all rs ought to be positive.

	1	2	3	4	5	6
Table 10.3: Correlation Matrix for Social Support Items						
1	---					
2	.415	---				
3	.378	.727	---			
4	.535	.436	.452	---		
5	.441	.426	.383	.650	---	
6	.417	.569	.525	.517	.396	---

All are positive, and none are so high that items appear redundant, so it makes sense to do further tests to confirm that the items are behaving as they should.

A correlation matrix can be displayed in a correlogram, as shown in Figure 10.4. Darker shades indicate stronger correlations. Note that the three variables grouped on the lower right, scale items 2, 3, and 6, show consistently high correlations. These are the items that were not reverse worded, so the graph is suggesting that the direction of wording affects the performance of items in this set.

Figure 10.4: A Correlogram for the Social Support Items

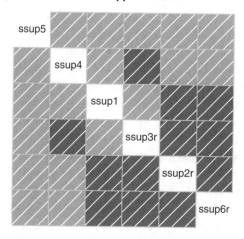

Graphic Displays for Interval or Ratio Data

The kind of data for which r is useful can be displayed in scatterplots. Figure 10.5 illustrates how the graph changes with r. On the upper left we have no linear association at all (r = .00). On the upper right we have a quite modest r (.30). It is followed by a moderate one (r = .60) and finally a strong one (r = .90).

Figure 10.5: Plots Illustrating Four Values of r

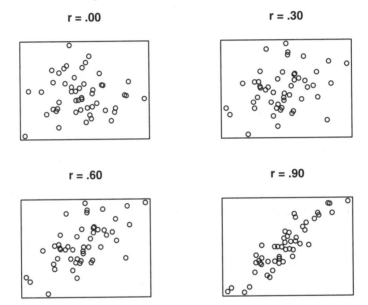

If we do not need to show the location of every point, but must indicate what is happening at each value of an independent variable, we can shift to a boxplot, as shown in Figure 10.6 for level of education and monthly income.

With interval or ratio data, we can also incorporate trend lines, but graphs including them will be left to the section on regression, for which trend lines are central.

In regression, we ordinarily hope to identify trends that can be represented by a straight line. In situations where we cannot summarize a relationship well with a straight line, and we have interval or ratio variables, we may choose to use a line or bar graph. The line graph lends itself very well to showing how average values of y change with x. It thus lends itself well to displaying changes in rates over time. In Figure 10.7 we see the Canadian female employment rate plotted against year. Since age of youngest child affects female employment considerably, separate lines are shown for different ages. (In this way we have shown the specifying effect of a third variable.)

Figure 10.6: Monthly Income by Years of Education

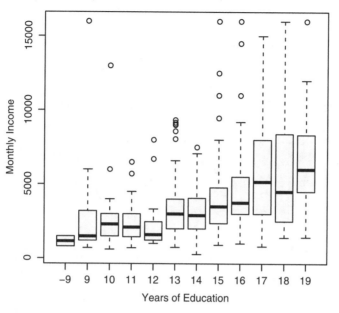

Figure 10.7: Female Participation Rate by Age of Youngest Child

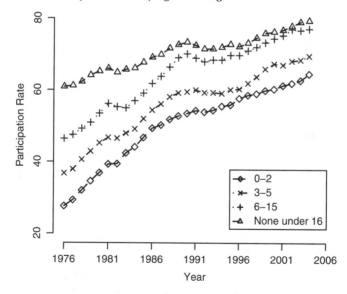

It is common to remove minor squiggles by creating moving averages. We calculate means using data from N consecutive points at a time, and plot the means at the points midway through the periods, as shown in Figure 10.8 with three-year averages.

Figure 10.8: Female Participation Rate by Age of Youngest Child (Three-Year Moving Averages)

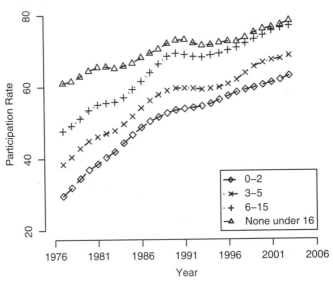

Sometimes a bar chart whose heights represent mean scores or proportions is a reasonable alternative to a line graph, as illustrated with Figure 10.9.

Figure 10.9: Mean Annual Household Income by Level of Education, Canada, 2007

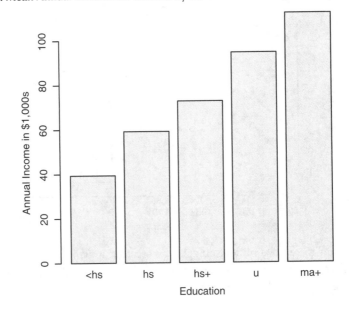

The level top of the bars can make it easier to estimate a value on the y-axis, but at the same time can break up what would otherwise be a smooth trend line.

If coloured, a bar graph will often have greater visual impact than a line graph, but will have a higher ink-to-information ratio. (Some writers suggest that graphs should make their points as elegantly as possible, and from this point of view a key measure of graphic style is the ink-to-information ratio.)

Table 10.4: Measures of Association by Level of Measurement

Basic Level of Measurement	Measure of Association	PRE?	Type of Error	
			Counted	Formula
Nominal	lambda	Yes	Cases not in modal category	$(\Sigma m_j - M) / (N - M)$
	Q	Yes	Pairs not in predicted direction	$(ad - bc) / (ad + bc)$
	odds ratio	No	N.A.	ad / bc
Ordinal	gamma	Yes	Pairs not in predicted direction	$(C - D) / (C + D)$
	d	Yes	Pairs not in predicted direction, plus ½ pairs tied on DV, but not IV	$(C - D) / (C + D + T_y)$
	rho	No	N.A.	As for r below, but using ranks rather than variable value
Interval-Ratio	r	No	N.A.	$Cov(x,y) / (SD(x)SD(y))$
	r^2	Yes	Squared errors in predicting one variable from another	square of the above

Summary

In this chapter we have seen the rationale for r, which is the most common measure of association for interval and ratio variables (and ordinal variables treated as interval). Although the rationale for the formula does not lead directly to a useful interpretation, two are available. First, r tells us how much of an SD of change in one variable is associated with an SD of change in the other. Second, if we square r it gives us the proportion of Variance in one variable that can be accounted for by the other.

Two variations on r are fairly common: rho (ρ), which works with the ranks of observations rather than their specific scores, and phi (φ), which is calculated for a 2×2 table.

The rs among a set of variables often appear in a correlation matrix, which can be represented in a correlogram. We may display the kind of data for which r may be calculated in a scatterplot, in a boxplot, or in a bar chart. A variation on the scatterplot can also be a good display for trends over time. Points representing individual time periods can be joined by point-to-point lines to clarify what has happened.

Review Questions on Pearson's r

1. In the formula for r, what is one function of the SDs in the denominator?

2. Why is the numerator of Pearson's r referred to as the "Covariance"?

3. In the formula for r, how is evidence of a positive association tallied up? Of a negative association?

4. Starting from a formula for r that does not use algebraic notation, show what happens when the variables are standardized.

5. What are two ways to interpret Pearson's r?

6. What is the difference between Spearman's ρ and Pearson's r?

7. What do we do before calculating rho if more than one case lies in a category?

8. Why might we find entries only in the lower triangle of a correlation matrix?

9. What is a scatterplot? What are some alternatives?

10. What is a moving average?

11. What are two advantages of a bar chart over a line graph? Two disadvantages?

Note

1. This formula is one typically used for populations rather than samples. In the sample version, N is replaced by (N – 1). The value of r is the same either way because the Ns or (N – 1)s above and below cancel.

Part IV
EXAMINING CROSSTABULATIONS

In a review of the state of statistics in 2000, Raftery (2002) pointed out that sociologists had led the way in providing methods for analysis of crosstabulations. The methods they have developed have often been used in some neighbouring disciplines. Often examination of two- and three-way tables is the best way to get clear on the links among variables. As well, in applied work clients often want us to obtain and explain crosstabulations for them.

We have seen several measures used to assess association in crosstabulations. We need also to seek out trends or patterns within them, to identify unusual cells within them, and to use graphs to represent them. Chapter 11 shows how we do these for two-way tables (those involving only two variables).

In Chapter 12, which deals with conditional tables, we will see what may happen to a relationship between two variables when we control for a third.

11

Two-Way Tables

Learning Objectives

In this chapter, you will learn

- how to read a crosstabulation;
- how to identify heavy (or light) cells;
- how to set up a crosstabulation for presentation; and
- some useful graphic methods of presenting cross-tabulated data.

Reading a Crosstabulation

Crosstabulations can provide multiple forms of information, which will be illustrated through Tables 11.1 and 11.2. In Table 11.1 we have two nominal variables: the dependent is the party voted for in the federal election of 2008, and the independent is region. The categories of the independent, region, are placed in the column headings, and percentages have been calculated separately for each region. The percentages sum to 100, or to a figure not quite equal to 100 because of rounding, for each region.

This is the typical way to set up a crosstabulation. When they are in this form, we see the effect of the independent variable by comparing across the table. If region affects party preference, then the percentages supporting the parties should differ across regions, and we should be able to see this by comparing the percentages backing a specific party across columns. For example, we can see that 20.8% of respondents from Quebec chose the Conservatives, but that fully 57.3% of those from the Prairies made this choice.

Table 11.1: Party Voted For by Region, 2008, in Percentages

Party	Region					
	Atlantic	Quebec	Ontario	Prairies	BC	Total
Bloc Québécois	0.0	40.5	0.0	0.0	0.0	9.3
Conservative	30.6	20.8	43.2	57.3	45.6	39.1
Green	9.5	3.1	6.6	8.0	8.1	6.6
Liberal	37.0	23.4	33.4	16.7	16.5	26.2
NDP	22.8	12.1	16.8	17.9	29.8	18.8
Total	99.9	99.9	100.0	99.9	100.0	100.0
N	359	644	912	436	467	2818

Crosstabulations provide various forms of useful information. Most obviously, they provide specific figures we can look up. If we want to know what proportion of Quebecers in this sample supported the Parti Québécois, we can look in the second column, headed "Quebec," and in the first row, "Bloc Québécois," to find the figure of 40.5%. As we have seen, we can compare specific figures as well. We can also look for a set of numbers to answer a question. We might, for example, ask which region was most favourable to each of the parties. In the case of the Bloc, the answer is obvious, since it runs no candidates outside Quebec. For the other parties, we see that Conservative support runs highest on the Prairies, at 57.3%; the Greens seem to be doing best in the Atlantic region, where they have 9.5% of the votes; the Liberals also peak in the Atlantic region, at 37.0%; and the NDP hits its maximum share of the vote, at 29.8%, in BC.

Table 11.2 involves two ordinal variables, education of wife and education of husband.

Table 11.2: Education of Husband by Education of Wife, in Percentages

Education of Husband	Education of Wife				
	< High school	High school	Post-secondary	University degree(s)	Total
< High school	53.8	19.5	6.8	6.4	13.5
High school	23.1	43.7	12.9	4.3	19.0
Post-secondary	20.5	32.2	67.8	25.5	51.9
University degree(s)	2.6	4.6	12.5	63.8	15.6
Total	100.0	100.0	100.0	100.0	100.0
N	39	87	264	47	437

In this table, we can again look up specific figures, or seek out those that answer a specific question. Now, though, because the independent variable is ordinal, we can look for trends. Consider the first line, which shows husbands with less than complete high school education. As the education of the wife rises, the percentages of men at this level drops from a high of 53.8% to a low of 6.4%. The higher the education of the wife, the lower the proportion of husbands with less than a high school diploma. We can also look for systematic rises and falls as we move across the table. On the second line, we see that men with high school diplomas have most commonly been chosen by women at the same level, and that the proportion of men at this level drops off as the wife's education either falls or rises. The third line shows a rise up to the post-secondary level, then a fall. The fourth line shows the highest figure in the final column: of university-educated women, 63.8% have husbands with degrees. Notice that the highest percentage in each row comes up with men and women who are in the same category. These results provide evidence of homogamy, or a tendency for like to marry like.

Heavy and Light Cells

In examining crosstabulations, we often try to identify cells that have far more cases, or perhaps far fewer, than might be expected. These are sought out for several reasons:

a) They may make it easier to interpret a table. It is simpler to focus on a small number of unusually heavy or light cells than to interpret them all, and we may find that most of what is interesting in the data is then dealt with. Let us consider the association plot in Figure 11.1. Heavy cells (where O > E) are represented above a line and shaded dark grey, while light cells (where O < E) are placed below the line and shaded light grey. The shaded rectangles are proportional to the difference O – E.

Figure 11.1: Party Preference by Region

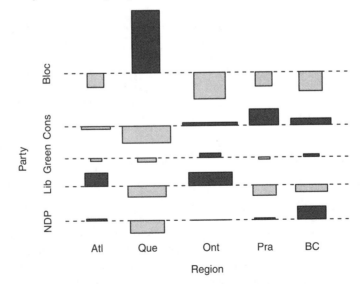

We see here the strong representation of the Bloc Québécois in Quebec, as well as the very strong showing of the Conservatives on the Prairies and their solid showing in BC. The Greens do about as well in one region as in another. The Liberals show strongly in the Atlantic region and in Ontario, and the NDP show strongly in BC. Plainly, if we can account for the parties' strong showings in these regions, we will have made sense of much of what the data have to tell us.

b) They may account for the relationship between two variables. If a measure of association shows a link, this may involve only a few heavy cells. If this is so, then an explanation of why they are heavy is an explanation of the association.

c) They may show us where a relationship is particularly strong (or weak). For example, sociologists have often examined tables showing occupation of son (or daughter) by occupation of father. Often the heaviest cell is that in which both fathers and their children are in the highest occupational category. This provides evidence that transmission of occupational standing is strongest at the top.

In looking for unusually heavy or light cells, we often go over the standardized residuals.

Standardized Residuals

Standardized residuals are a by-product of calculating chi-square. Recall that in calculating chi-square we take the sum, across all cells, of

$$(O - E)^2 / E$$

The square root of this,

$$\pm [(O - E)^2 / E]^{.5} ,$$

is called the standardized residual (SR). One can be obtained for every cell in a table.

The sign given to a standardized residual depends on whether the observed cell value (O) is greater or smaller than the expected value (E). (Recall that the latter is the value expected in a sample if the row and column variables are unrelated in the population from which the sample was drawn.) Where the cell is heavy, the standardized residual will be positive; where the cell is light, it will be negative.

If we draw a great many samples, and if the row and column variables are unrelated in the population, the standardized residuals from our many samples (if the Ns are of a reasonable size) will take on a shape very like the normal distribution. Recall that in a normal distribution, 95% of the cases lie within two standard deviations of the mean.[1] Thus the standardized residual for a given cell should be less than approximately ± 2 in 95 samples out of 100. So, if we get a standardized residual greater than ± 2, we are likely to look at the cell to see what it may be telling us.[2]

The rule of thumb that we take seriously standardized residuals with absolute values of 2 or more is subject to restrictions, though:

a) If E is small, standardized residuals do not have a normal distribution. For tables considered here, Es will not be an issue.

b) If the sample size is large enough, many of the standardized residuals are apt to exceed 2 in absolute value. Differences between O and E that mean little may then be large numerically. With a large sample, we tend to look only at the greatest of the residuals.

Visual displays of standardized residuals are available, as in Figure 11.2, which provides values for the standardized residuals for each cell. In the legend for the graph, they are called "Pearson residuals" since they are by-products of chi-square, whose inventor was Karl Pearson. The SRs greater than ± 2 are represented by dotted patterns: black dots represent positive SRs, and grey dots represent negative SRs. Larger dots tell us the SRs are > ± 4.

Figure 11.2 is a "mosaic plot" showing education of husband by education of wife. In this plot, each cell is represented by a rectangle whose area is proportional to the number of cases in the cell. The SRs themselves are indicated by the patterns and the shading. Solid grey cells lie below ± 2.0. Heavier cells are shown in black-and-grey dotted patterns; lighter cells are shown in grey-and-white dotted patterns. Cells with values > 4.0 are shown in the black-and-grey dotted pattern with the larger dots.

Figure 11.2: Husband's Education by Wife's Education

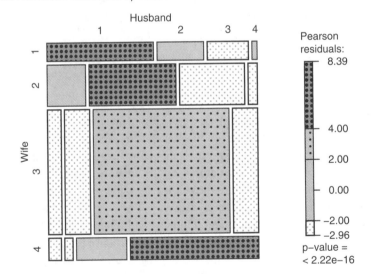

Levels of education are represented by numbers in this graph:

1 – less than complete secondary;

2 – complete secondary;

3 – post-secondary, but no university degree; and

4 – university degree or degrees.

The cells along the diagonal (top left to lower right) all have SRs > 2.0, and the cells on the upper left and lower right come in above 4.0. The SRs confirm our initial impression of homogamy here.

Further, we have SRs < −2.0 in four of the lower left cells and in three of the upper right cells. The combination of heavy cells along the main diagonal and light cells in the lower left and upper right corners suggest a solid positive correlation.

If we wish to examine the SRs in more precise detail, readily available statistical software will provide them. Table 11.3 presents them for the plot we have been examining.

Table 11.3: Standardized Residuals for Education of Husband by Education of Wife

Education of Husband	Education of Wife			
	< High school	High school	Post-secondary	University degree(s)
< High school	6.9	1.5	−3.0	−1.3
High school	.6	5.3	−2.3	−2.3
Post-secondary	−2.7	−2.6	3.6	−2.5
University degree(s)	−2.1	−2.6	−1.3	8.4

Besides confirming the general impressions we have obtained graphically, the specific figures allow us to see that the most extreme figures are obtained on the upper left and lower right, where the SRs are, respectively, 6.9 and 8.4. The weight of these cells adds to our impression of a positive correlation between these two variables.

It may be, then, that it would be better to look at the overall pattern here, seeking an explanation for the correlation between husbands and wives rather than an explanation for a set of heavy cells. Assessing this will require us to look again at the crosstabulation, which reappears in Table 11.4.

Table 11.4: Education of Husband by Education of Wife, in Percentages

Education of Husband	Education of Wife				
	< High school	High school	Post-secondary	University degree(s)	Total
< High school	53.8	19.5	6.8	6.4	13.5
High school	23.1	43.7	12.9	4.3	19.0
Post-secondary	20.5	32.2	67.8	25.5	51.9
University degree(s)	2.6	4.6	12.5	63.8	15.6
Total	100.0	100.0	100.0	100.0	100.0
N	39	87	264	47	437

Lambda = .195, p < .001; d = .511, p < .001, gamma = .683, p < .001

We can immediately see that the largest percentage for each row falls on the main diagonal, exceeding 50% for three of the four cells and 60% for two. These results suggest that if we can explain the tendency for like to marry like, we have accounted for much of what is happening in the table. However, this tendency to homogamy could be viewed as part of an overall tendency for husbands and wives to correlate. Not only are the highest percentages found where the husband's education matches the wife's, but there is only one minor exception to another pattern. Apart from one cell in the fourth row, in each row the percentages fall regularly as we move away from the homogamous cell. This pattern implies a clear positive correlation.

Since there is no clear dependent variable, we might prefer symmetric gamma over asymmetric d here, and gamma comes in at .683. But d, which penalizes the column variable for tied scores on the row variable, still comes in at .511, far above the nominal level measure lambda at .195. The moral is that we have a correlation that goes beyond the strict homogamy represented by the heavy cells on the diagonal. We are not barred from trying to explain why they are so heavy, but we must also try to understand the more general tendency for the education of spouses to rise and fall together.

Setting Up a Crosstabulation for Presentation

Each discipline, publishing house, and often an individual journal will have its own guidelines for style. Nonetheless, a certain set of features is expected in a crosstabulation. Here we will see how to create a table including them.

Before beginning, we need to choose a monospace (or fixed-width) font, one in which each character is the same width. Numbers in the table must be aligned, and alignment is extremely difficult if the characters are of different widths. Some common monospace fonts are Andale Mono, Bitstream Vera Sans Mono, Courier, Courier New, DejaVu Sans Mono, FreeMono, Lucida Console, and Monaco. We can choose from among those available to us on grounds of clarity and appearance.

We begin with a table title and headings for the columns. The title ordinarily begins with the name of the dependent variable, followed by the word "by," then the name of the independent variable. It may be necessary to add something further to explain some point not clarified in the text or needing emphasis: where the data were gathered, that the table entries are percentages, or that the table has a specific number in a sequence of tables. The full title should be centred over the table, and printed in boldface or italics. Here we use boldface in the title of our illustrative table.

Table 11.5: Party Voted For by Region, 2008

Below the title, we often insert the name of the independent variable, centred over the column headings, and printed in bold (or italics, if they have been used for the title).

Table 11.5: Party Voted For by Region, 2008

Region

Below the title and the name of the independent variable, column headings appear. On the left, we place the name of the dependent variable. Moving to the right, we name the categories of the IV, and then indicate that the final column on the right includes all cases in the sample. Everything here ordinarily appears in boldface or, if the title is in italics, the headings will be as well. Following these guidelines, and consistently using boldface, we obtain the following:

Table 11.5: Party Voted For by Region, 2008

			Region			
Party	**Atlantic**	**Quebec**	**Ontario**	**Prairies**	**BC**	**Total**

Next, on the left, in the area below the name of the dependent variable, we place the row labels, the names of the categories of the dependent variable. Unlike the categories of the IV, these are not bolded (or italicized). Below them, we place the label "Total," so the line with column totals will be clearly identifiable. We obtain

Table 11.5: Party Voted For by Region, 2008

			Region			
Party	**Atlantic**	**Quebec**	**Ontario**	**Prairies**	**BC**	**Total**
Bloc Québécois						
Conservative						
Green						
Liberal						
NDP						
Total						

Now we have to fill in the "body" of the table. This is the area to the right of the row labels (but not including the area beneath the column heading "Total"). Since the table has not been percentaged, we just fill in the number of cases found in each cell.

The numbers must be centred beneath the column headings, and figures above and below one another must be precisely aligned. Since we have no decimal points, we align the rightmost digits within each column. When we have done this, the table appears as shown:

Table 11.5: Party Voted For by Region, 2008

			Region			
Party	**Atlantic**	**Quebec**	**Ontario**	**Prairies**	**BC**	**Total**
Bloc Québécois	0	261	0	0	0	
Conservative	110	134	394	250	213	
Green	34	20	60	35	38	
Liberal	133	151	305	73	77	
NDP	82	78	153	78	139	
Total						

Next we add the "marginal totals" in the bottom row and rightmost column, representing the row and column totals.

Table 11.5: Party Voted For by Region, 2008

| | Region | | | | | |
Party	Atlantic	Quebec	Ontario	Prairies	BC	Total
Bloc Québécois	0	261	0	0	0	261
Conservative	110	134	394	250	213	1101
Green	34	20	60	35	38	187
Liberal	133	151	305	73	77	739
NDP	82	78	153	78	139	530
Total	359	644	912	436	467	2818

If we wanted to display only the cell counts, the table could stand in its present form. Typically, though, we want to display some statistics. Since this is a nominal-by-nominal table, the possibilities are limited, but we will include those we have considered. We may place statistics below the table, visually separated from the column totals, or we may place a line beneath the totals, and present our statistics below the line. Here we will follow the simpler course, omitting the line.

Table 11.5: Party Voted For by Region, 2008

| | Region | | | | | |
Party	Atlantic	Quebec	Ontario	Prairies	BC	Total
Bloc Québécois	0	261	0	0	0	261
Conservative	110	134	394	250	213	1101
Green	34	20	60	35	38	187
Liberal	133	151	305	73	77	739
NDP	82	78	153	78	139	530
Total	359	644	912	436	467	2818

X^2(16 df) = 1126.572, $p < .001$, lambda = .087, $p < .001$

Percentaged tables require some indication of the Ns on which percentages are based. There are many ways to include them. Here we will present them in a row beneath the column totals. Similarly, there are many ways to indicate that cell entries are percentages. Here we declare this in the title.

The sum of the percentages is sometimes 99.9 because of rounding. Small departures from 100 often occur, and usually are not remarked on, although you will see footnotes to the effect that "percentages do not sum to 100 due to rounding."

Table 11.6: Party Voted For by Region, 2008, in Percentages

Party	Atlantic	Quebec	Ontario	Prairies	BC	Total
			Region			
Bloc Québécois	0.0	40.5	0.0	0.0	0.0	9.3
Conservative	30.6	20.8	43.2	57.3	45.6	39.1
Green	9.5	3.1	6.6	8.0	8.1	6.6
Liberal	37.0	23.4	33.4	16.7	16.5	26.2
NDP	22.8	12.1	16.8	17.9	29.8	18.8
Total	99.9	99.9	100.0	99.9	100.0	100.0
N	359	644	912	436	467	2818

X^2(16 df) = 1126.572, p < .001, lambda = .087, p < .001

Note that we have percentaged down the columns rather than across the rows. When the IV is the column variable, and the DV the row variable, this is standard, since we want to show the impact of the IV. If it has an effect, the DV must show different values when we move from one category of the IV to another. In this table, if region makes a difference, we must be able to see it when we move from Atlantic to Quebec, or between some other pair of categories. Specifically, the percentage who prefer particular parties must change as we move from one region to another. If we have percentaged down the columns, we can see how the figures change as we move across, from one category of the IV to another. More generally, when the IV is the column variable, *the rule is to percentage down, and to compare across*. In this way, we can directly examine the impact of the IV.

Note as well that we have placed only one digit to the right of the decimal point. Depending on the precision required, or the practices of a journal, more may be needed. Since these require no changes to the basic format of the table, an example with further digits will not be provided here.

Further Graphic Methods of Clarifying a Crosstabulation

This section deals with three types of plot: association plots, side-by-side bar charts, and stacked bar charts. Each enables us to emphasize specific aspects of the data.

We have seen examples of how graphs can display standardized residuals, through association plots and mosaic plots. If we do not need to show the residuals, but merely to draw attention to cells that are heavy or light, we can use a simple association plot, without the residuals included. Figure 11.3 shows how custodial mothers' stance on visitation was affected by their feeling toward their ex-husbands.

Recall that, in this form of plotting, heavy cells, here shaded dark grey, rest above a line, while light cells, here shaded light grey, rest below it. The size of the rectangles is proportional to the difference between O and E. This graph shows that when the mother's view of her ex is negative, she is more likely to discourage visiting than when it is not. Conversely, when her view is not negative, she is more likely to take another position, perhaps that it is a good thing, or that visitation is up to the father and the child, or perhaps that it is fine as long as some conditions are met.

Figure 11.3: Stance on Visitation by View of Ex-husband

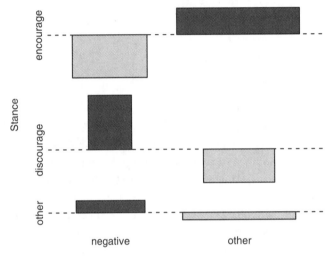

If we do not need to display figures, but we want to compare the frequency of several categories, we can use bar charts. One form of bar chart, illustrated in Figure 11.4, places the categories beside one another.

Here it is clear that the number of single-parent families in each category is much higher in 2006 than in 1966. It is also clear that the proportions who are divorced or never married is much higher in 2006. The same points can be made in a stacked bar chart, shown as Figure 11.5.

The second version of the graph makes the sheer increase in single parenthood between 1966 and 2006 more obvious because there are only two columns to compare. The simplicity of the comparison also draws attention to the thinness of the bars with a dotted pattern (for divorced parents) and grey bars (for never married) on the left as compared to their thickness on the right. The author prefers this version, but there is no fixed rule about when to stack the categories and when to place them side by side. We have to decide in terms of what we want to emphasize and the appearance of the graph.

An issue we do have to confront is how complex the graph should be. As more pieces of information are added, it becomes more difficult to convey a message quickly.

Figure 11.4: Single-Parent Families by Marital Status of Head, 1966 and 2006, in 1,000s

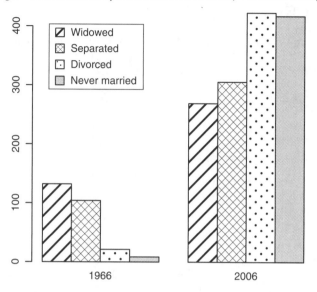

Figure 11.5: Single-Parent Families by Marital Status of Head, 1966 and 2006, in 1,000s

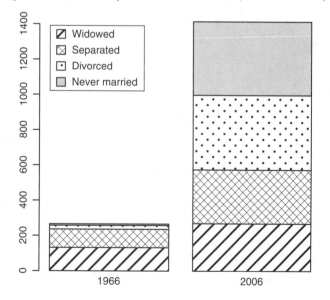

Consider Figure 11.6, which shows education of husband by education of wife in a side-by-side bar chart.

Figure 11.6: A Side-by-Side Bar Chart for Education of Husband by Education of Wife

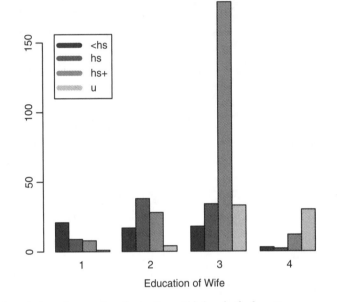

The numbers below the graph refer to the wife's level of education:

1 = less than complete high school;

2 = high school;

3 = post-secondary; and

4 = university degree(s).

In this chart there are 16 pieces of information. At first glance, it is obvious that when wives are in category 3, post-secondary education, there are a great many husbands in the same category. After noticing that, many people would have to study the graph carefully to see what else it had to say. They might do as well to examine a table like 11.4, which presents the same data in a different format.

When we want to use a graph, but we also want its message to be straightforward, one solution is to focus on one part of the data at a time. For example, Figure 11.7 shows level of education for husbands who have university-educated wives.

Here we readily see that the number of husbands rises with the level of education, with one exception, and peaks where husbands are at the same level as their wives.[3]

Figure 11.7: Education of Husbands, for Wives with University Degrees

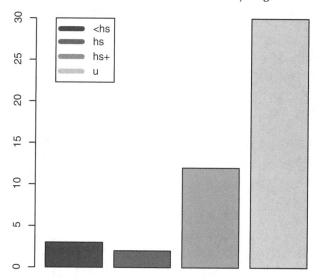

Summary

This chapter has treated the most important ways in which we use two-way crosstabulations. We can extract information directly from them, perhaps the percentage in a specific cell or the difference between two columns, or perhaps trends as we move across the table. We have seen how to search for unusual cells, through the standardized residuals, and we have seen how these can be represented graphically. Finally, we have seen how to set up a two-way crosstabulation for presentation.

Sometimes, though, the link between variables that appears in a two-way table changes considerably when we break the sample down into subcategories and produce new tables for each. Chapter 12 will address this question.

Review Questions on Heavy and Light Cells and Graphic Methods

1. What is a mosaic plot? Why are the rectangles in the plot of different sizes? Why do we care about the "Pearson residuals"?

2. Why are some cells shaded or patterned differently? Why might we be interested in a cell that is particularly heavy (dark) or particularly light?

3. If we wanted to percentage the table showing vote by region, discussed above, in which direction would we do this? How would we then interpret the differences in percentages?

4. What measure do we typically use to identify heavy cells? How is it related to chi-square? What values of the measure are we typically interested in?

5. What is an association plot? What is the difference between rectangles above a line and those below?

6. In Figure 11.2, what do the cells tell us, through their shading and patterning, about how a person's choice of spouse is linked to her or his level education?

Notes

1. Recall that we calculate a z-score, or standard score, by taking $(x_i - \bar{x}) / SD(x)$. Given this formula, we can see why the term "standardized residual" is used. If the rows and columns are independent, the distribution of standardized residuals for a given cell will have zero as its mean, and the standard deviation of the distribution will be the square root of E. We take a residual $(O - E)$ and subtract the mean of the residuals we would get from many samples, which is 0, then divide through by the SD of the distribution, which is the square root of E. The result, $(O - E) / \sqrt{E}$ has been standardized in a way analogous to the construction of a z-score, commonly also called a standard score.

2. To be correct technically, we should look at only as many cells as there are degrees of freedom in the table. In practice we tend to use standardized residuals heuristic-ally, that is as aids in locating cells of substantive interest rather than as the basis for technically correct inferences as to whether cell residuals might have come up by chance.

3. Another option, coming from the field of statistical graphics, is the trellis graph, which places several graphs adjacent to one another in a common frame. These graphs, while very efficient, have not yet become common in the social sciences.

12

Conditional Tables

Learning Objectives

In this chapter, you will learn how introducing a third variable can alter the apparent association between two others. In particular, you will see

- how relationships may differ between subgroups;
- how causal chains may become apparent;
- how an association due to a common cause may come up; and
- how the apparent direction of an association may change.

The Columbia Approach

To check the ways an association might change when a third variable was controlled, the Columbia school proposed a series of tests. Each required the original table to be recreated for subgroups—for example, for males and females. Because tables were created for people who met specific conditions, they could (unsurprisingly) be referred to as conditional tables. They are also called "partial tables."

Each conditional table includes only people for whom the value of a third variable, called a "test factor" is fixed. Since it is fixed, the "test factor" cannot affect the link between the other two within that table. Between tables, on the other hand, the third variable changes. If the link between the others changes as well, we have to ask what the third variable had to do with the change.

This strategy has a clear practical limitation. If we break the data down in many ways simultaneously, we will have few cases left in some subtables. Suppose we break down a sample by sex, by nativity (Canada vs. Other), by level of education and province of residence. Without huge samples we will have too few cases to be helpful in some of the subtables that will result.

One way to try to get around this is to collapse the test factors. For example, education might be dichotomized as "high school incomplete" and "high school complete." This would sometimes make sense, but those in the subtables would no longer be identical in their education, so control for its effects would be incomplete.

If we have large samples, or we do not need to divide them several ways at once, introducing a third variable can assist us in sorting out the links between two initial variables. If, for example, a relationship appears much stronger for males than for females, the original results can be viewed as a kind of average of what is true for the two sexes. Technically, this result is "specification" and sex of respondent is a "specifier"—a variable that helps us to specify more precisely what is going on by distinguishing among subgroups.

Specification

The term "specification" has a long history in sociology, but in other fields, and in sociology as well, you will see another term, "moderation," used to mean the same thing. If the relationship between two variables changes when the value of a third changes, the third may be called a "moderator." This usage is standard in some disciplines.

Specification appears in work by Hamilton and Pinard (1976) on support for the Parti Québécois (PQ). We can view their results either in a plot, or in tables adapted from the original. The plot, shown in Figure 12.1, is referred to as a "doubledecker," because of the two variables identified in the layers at the bottom.

Here, location (Montreal, abbreviated "Mon," or elsewhere) and education (elementary or more) are distinguished.

Note that the width of the bars reflects the size of the subgroups. The smallest consists of those with elementary education living in Montreal, and the largest of those with more than elementary living elsewhere in Quebec. On the vertical axis, party preference is indicated, with the PQ in dark grey and other parties in light grey. PQ support is highest in Montreal for both educational categories, but the difference between the metropolis and the rest of the province appears greatest for those with elementary education.

The two-way tables shown in Table 12.1 suggest that education raised PQ support, as did living in Montreal. For each table, Q is substantial (at −.463 for the education table and .591 for the place of residence table).

Figure 12.1: PQ Vote by Residence and Education

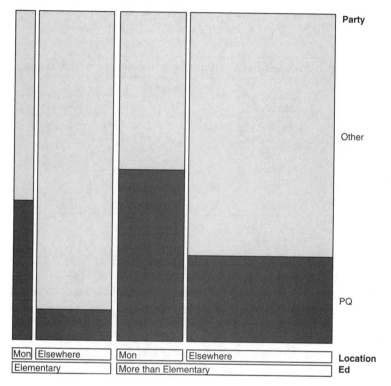

Table 12.1a: Party Support by Education, in Percentages

Party	Education		
	0–7 Years	8+ Years	All Cases
PQ	17.8	32.9	28.4
Other	82.2	67.1	71.6
Total	100.0	100.0	100.0
N	256	556	812

Q = −.463

Table 12.1b: Party Support by Residence, in Percentages

Party	Residence		
	Montreal	Elsewhere	All Cases
PQ	49.8	20.3	28.4
Other	50.2	79.7	71.6
Total	100.0	100.0	100.0
N	225	587	812

Q = .591

However, the three-way Table 12.2 shows that the impact of education was greatly affected by living in Montreal. We saw in the doubledecker that education had a greater effect on those living there. The three-way table makes the difference precise.

	Table 12.2: Party Support by Education by Residence, in Percentages					
	Residence					
	Montreal			**Elsewhere**		
Party	**0–7 Yrs**	**8+ Yrs**	**All Cases**	**0–7 Yrs**	**8+ Yrs**	**All Cases**
PQ	42.3	52.0	49.8	9.3	26.1	20.3
Other	57.7	48.0	50.2	90.7	73.9	79.7
Total	100.0	100.0	100.0	100.0	100.0	100.0
N	52	173	225	204	383	587
	$Q = -.193$			$Q = -.550$		

As noted above, Q is not affected by marginal totals, so it is well suited to dealing with conditional tables. For those in Montreal, the absolute value of Q moves sharply toward zero from its value in the two-way table (−.463) to −.193, but for those elsewhere Q becomes more impressive, going to −.550. Place of residence thus "specifies" the effect of education. We can write the specification effect this way:

In Montreal, Education → Party Preference

Elsewhere, Education → Party Preference

Expressing the causal effect in this way just says that education's impact on party preference differs by place of residence. In this case it is strongest outside Montreal.

Causal Chains

Social science has given great attention to causal chains. Chapter 15, on path analysis, will be devoted entirely to one way of looking at them. Here we will see how conditional tables may help us to sort out the workings of such chains.

A widely used example comes from Berkeley in the 1970s (Bickle, 1975). It appeared that females applying for graduate work might have been unfairly treated. Only 30% of females, against 44% of males, were accepted. The basic data are found in Table 12.3.

To sort out what was happening, admissions were examined by program. Key results are presented in Figure 12.2 in a doubledecker, in which six of the largest programs at Berkeley are displayed. Recall that in a doubledecker the width of the bar is proportional to the size of the category, and note that the proportion of males is much higher for programs A and B than for any of the others. Notice that these are also the programs which accepted the highest proportions of applicants, whether male or female. Also, the difference in the proportion admitted is most striking for program A, where females were more often accepted.

Table 12.3: Admission by Sex		
	Male	Female
Admitted	44%	30%
Rejected	56%	70%
Total	100%	100%
N	8,442	4,321

Q = .294

Figure 12.2: Admission by Sex by Program

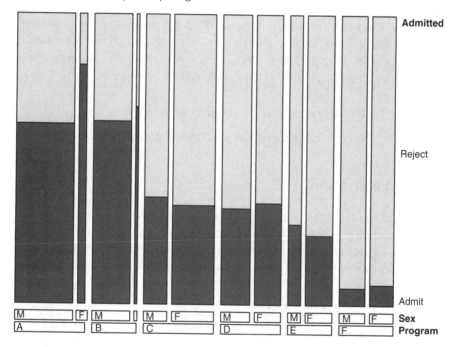

Here are the percentages of males who applied, by program, and the overall percentage accepted.

	A	B	C	D	E	F
% Male	88.4	95.7	34.5	52.7	32.6	52.2
% Accepted	64.4	63.2	35.1	35.3	25.1	6.5

By far the highest proportions of applicants who were male were found at the two departments with the highest admission rates (A and B).

In short, the higher overall rate for males resulted from their applying for programs to which admission was easiest. If we want a single measure of the change in the apparent effect of sex, we can compare our original Q, at .294, with "partial Q," calculated as

$$\frac{\Sigma\, ad - \Sigma\, bc}{\Sigma\, ad + \Sigma\, bc},$$

where the Σs imply that we will add across the conditional tables. If we calculate Q in this way, we take out the impact of differences among programs because each ad and each bc will come from a table representing only one program.

Here, partial Q = −.060, a value opposite in sign, and only about a fifth as large as the original. Without controlling for our "intervening variable" we had a seriously inadequate picture.

In the end, it appears that there is no broad difference in admission rates between the sexes. It is only in program A that admission rates are strikingly different. If we recalculate partial Q, leaving out program A, it comes out at a very modest .016. The question we are left with is not a general one, but rather the specific issue of why admission rates to program A differed by sex.[1]

Looking at the broader picture, the causal chain can be written

Sex > Program Applied for > Admission

Spurious Association

Often a link between two variables results, at least in part, from their having a common cause. Most often, only part of their association comes about for this reason, and that will be the case for the example. Table 12.4 shows a link between whether the respondents acknowledged at least one session of binge drinking in the previous year and monthly household income.

Table 12.4: Binge Drinking by Monthly Income, in Percentages			
Binge	**<2,400**	**2,400+**	**Total**
No	73.6	85.0	77.9
Yes	26.4	15.0	22.1
Total	100.0	100.0	100.0
N	617	379	996

Q = −.340

For these data, Q = −.340, indicating that ad < cd. Here, the negative association indicates that lower incomes are associated with more binge drinking, and higher incomes with less.

In the same data set, single parenthood is associated with more binge drinking, and partnered parenthood with less. This is indicated by the negative value of Q (−322). Table 12.5 provides the data.

Table 12.5: Binge Drinking by Parental Status, in Percentages			
Binge	**Partnered Parent**	**Single Parent**	**Total**
No	82.3	70.5	77.9
Yes	17.7	29.5	22.1
Total	100.0	100.0	100.0
N	623	373	996

Q = .322

Single parenthood can be viewed as a cause of low income, relative to partnered parent-hood, since a household with only one potential earner is at a disadvantage, and parental status appears to affect binge drinking. We shall see the extent to which the link between income and drinking results from parental status. Table 12.6 provides the two conditional tables we need to check this.

Table 12.6: Binge Drinking by Monthly Income by Parental Status, in Percentages						
	Partnered Parent			**Single Parent**		
Binge	**<2,400**	**2,400+**	**Total**	**<2,400**	**2,400+**	**Total**
No	77.5	86.2	77.9	70.4	71.9	70.5
Yes	22.5	13.8	22.1	29.6	28.1	29.5
Total	100.0	100.0	100.0	100.0	100.0	100.0
N	274	347	623	341	32	373

Q = −.287   Q = −.036

For partnered parents, Q has moved modestly toward zero, from the original −.340 to −.287, but for single parents Q has become trivial at −.036. That is, income has essentially no effect for single parents, whose higher levels of binge drinking must be accounted for otherwise. If we calculate partial Q, as explained above, it comes to −.253, suggesting that a bit over a quarter of the apparent link between binge drinking and household income can be accounted for by parental status ((.340 − .253) / .340 = .256).

The difference between single and partnered parents in a way resembles a form of speci-fication. The effect of income clearly differs for the two family types. The reason we say that we have found a partially spurious association rather than simply a form of specification is that parental status can be seen as a cause of household income levels.

Distortion

An example of distortion, in which the original table shows an effect in the opposite direction to what we obtain after controlling for a third variable, comes from Appleton et al. (1996). In 1972–74 women aged 18–64 were interviewed, and 20 years later they were followed up to see how their health had turned out. Data from the follow-up showed that those who reported being smokers in the early 1970s were a bit less likely to have died in the meantime. A two-way table, shown as Table 12.7, can be created based on their data.

Table 12.7: Mortality by Smoking Status		
	Smokers	**Non-smokers**
Died	22.1	24.9
Alive	77.9	75.1
Total	100.0	100.0
N	569	568

Q = −.065

Given the way the variables have been coded, Q is negative here, for a table in which mortality is lower for smokers. We must watch for a sign change when an additional variable is introduced.

A third variable we can count on to help explain mortality is age. Its impact, with that of smoking, is shown in the double-decker in Figure 12.3.

Figure 12.3: Mortality by Smoking and Age

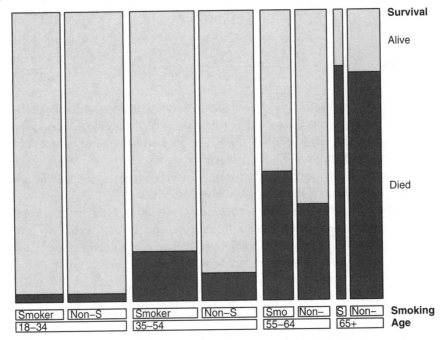

For all age categories but the first, smokers appear more likely to have died. Further, the proportion of smokers is clearly least in the highest age category. Age appears to be a "lurking variable" that must be controlled if we wish to understand the data.

The results of introducing age as a control variable are shown in Table 12.8. For simplicity, we show only the percentage who died, but of course the Qs below the conditional tables have been calculated from the full 2 × 2 tables.

Table 12.8: Mortality by Smoking Status and Age at Initial Interview

	Age at Initial Interview							
	18–34		35–54		55–64		65–74	
	Smoker	Non-Smoker	Smoker	Non-Smoker	Smoker	Non-Smoker	Smoker	Non-Smoker
% Died	2.8	2.7	17.2	9.5	44.3	33.1	80.6	78.3
N	179	219	239	199	115	121	36	129
	Q = .019		Q = .325		Q = .235		Q = .069	

Unsurprisingly, mortality rises with age, for both smokers and non-smokers. Being alert to the possibility of distortion, we note also that the Qs have changed sign, from negative to positive, implying that smokers have higher death rates now that we have controlled for age. The difference between smokers and non-smokers can be seen both in Q and in the percentages, which are higher for smokers in each age group.

For a single measure of the change in the apparent effect of smoking, we can compare our original Q, at −.065, with "partial Q," calculated as explained above. Here partial Q = .260, a value opposite in sign, and four times greater in magnitude than the original. Without controlling for our "lurking variable" we had a seriously distorted picture.

However, there is more than a switch of direction here. We see as well a form of specification, in that the Qs are quite different for different age groups. Those aged 18–34 at the beginning of the study would have been, on average, about 27 at their first interview, and 47 at the 20-year follow-up. Mortality would be expected to be low up to this age, whether a woman smoked or not, so we ought not to expect a great difference between smokers and others. Similarly, those aged 65–74 would have been, on average, just under 70 at the first interview, and if alive would have reached 90 by the follow-up. We would expect women in this age group to have very high mortality whether they smoked or not, so we ought not to expect a great difference between smokers and non-smokers.

It would be in the middle categories, for those aged 35–64 at the beginning of the study, where mortality would be high enough to show a clear difference, and low enough for it not to be obscured by the overwhelming impact of age. We find just this result, with the lowest Qs (.019 and .069) for the youngest and the oldest groups, and the highest Qs (.325 and .235) for those between.

In these data both age and smoking affect mortality, and age affects smoking. The causal links can be diagrammed as shown in Figure 12.4. By placing signs in the diagram, we indicate that age has a direct, positive effect on mortality, as well as a negative effect on smoking, which in turn causes mortality to rise.

Figure 12.4: Effects of Age and Smoking on Mortality

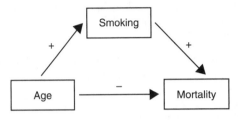

Conditional Probabilities

Without using the expression, this chapter has dealt with what are called conditional probabilities. These are just probabilities that apply under specified conditions. Using probabilities rather than percentages, part of the data relating age and smoking to mortality might have appeared as shown in Table 12.9.

	35–54		55–64	
	Non-smoker	**Smoker**	**Non-smoker**	**Smoker**
P(Died)	.172	.095	.443	.331
N	239	199	115	121

Table 12.9: Probability of Death by Smoking Status and Age

When we see P followed by the name of an outcome in brackets, we are being given the probability of that outcome. Here the percentage who had died, given in the full table, has been replaced by the probability of having died. Otherwise nothing has changed.

Conditional probabilities appear in the social science literature, but if we understand conditional tables, the corresponding probabilities will give us little trouble.

An Illustration with Polytomies

So far we have looked at associations among dichotomous variables. These are of interest in their own right, but we should also look at something a bit more complex. Table 12.10 links contraceptive practices to social class. In this table, socio-economic categories are coded so that 1 is at the top and 7 is at the bottom.

Two clear trends are present. In the first row, we see that the percentage using the most effective methods declines consistently as socio-economic position declines. In the fourth row, the proportion who object on principle rises. In the other rows, we see weaker trends. There are no reversals in the percentages as we move across the table, but cells are tied, and the differences among the cells are not as great as those for the first and fourth rows.

Table 12.10: Contraceptive Practice by Socio-economic Category, in Percentages

Practice	Socio-economic Category					
	1–2	3–4	5	6	7	All Cases
Pill, IUD, sterilization	75.0	71.1	67.0	64.9	47.8	65.3
Condom, diaphragm, chemical methods	10.7	6.0	7.0	7.0	5.1	6.7
Rhythm, withdrawal	5.4	5.9	7.0	7.0	11.5	7.3
None—objection on principle	7.7	8.5	10.0	12.1	24.1	11.7
None—other reasons	6.3	8.5	9.0	8.9	11.1	9.0
Total	100.1	100.0	100.0	99.9	100.0	100.0
N	112	682	1,011	838	370	3,013

Now we should compare those born in North America with those born elsewhere.

For the North American-born, only one cell, in the lower right, has a standardized residual of over 2.00, and the overall association would have arisen through sampling fluctuation if the rows and columns were independent 18 times out of 100. Since the association plot for the North American–born reveals nothing of interest, it has been omitted, but the plot for those born elsewhere, which does show differences by class position, is presented as Figure 12.5.

Figure 12.5: Contraception by Socio-economic Category for Those Born Outside North America

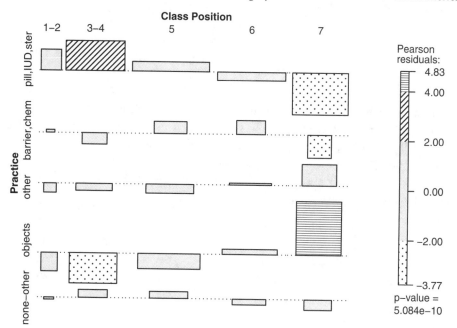

For those born outside North America, there are five highlighted cells, including three in the rightmost column. The cell in the upper right, representing use of the most effective methods of contraception, is identified as light, as is the one below it, representing use of the condom, diaphragm, and chemical methods. Balancing these light cells, there is one very heavy cell, which represents using no method because of objections on principle.

What we see here might be written as follows:

Born in N.A. → no demonstrable association

Born elsewhere → specific cells affected

Having seen this, we might wish to examine the tables themselves in more detail. The one for the North American–born will be omitted, since not much happens in it, but Table 12.11 provides figures for those born elsewhere.

Table 12.11: Contraceptive Practice by Socio-economic Category, for Those Born Outside North America, in Percentages

Practice	Socio-economic Category					
	1–2	3–4	5	6	7	All Cases
Pill, IUD, sterilization	65.2	52.9	43.3	37.5	22.0	40.5
Condom, diaphragm, chemical methods	8.7	5.3	8.8	9.4	3.0	7.3
Rhythm, withdrawal	8.7	13.8	13.7	16.0	21.4	15.6
None—objection on principle	8.7	16.4	22.6	28.1	45.8	26.6
None—other reasons	8.7	11.6	11.6	9.0	7.7	10.2
Total	100.0	100.0	100.0	100.0	100.0	100.0
N	23	189	328	256	168	964

In the top row, the downward trend from left to right is sharply defined. An upward trend is clearly present in the fourth row. With a minor reversal, another trend is visible in the third. If we compare the trends here with those in the table for the full sample displayed earlier, we discover that the trends here are much more pronounced. In fact, the trends in the sample are essentially created by those born outside North America, again illustrating the need to examine two-way tables by third variables.

Summary

In this chapter, we have seen illustrations of how introducing a third variable can help us to understand the link between an initial pair of variables. We have discussed

- specification, in which the link between two variables differs between subgroups defined by the values of a third;

- causal chains, in which the initial link between two variables is created when one variable influences another, which in turn affects the third;

- spurious association, in which the apparent link between two variables is due to a common cause; and

- distortion, in which the direction of an apparent effect reverses when we introduce a third variable.

We have primarily worked with 2 × 2 tables, but we have seen that introducing a third variable can sometimes greatly increase our understanding of a more complex table.

Review Questions on Conditional Tables

1. What are conditional tables? What is another name for them?

2. How do conditional tables "control for" third variables?

3. What, for the Columbia School, was a "test factor"?

4. What is a practical problem in breaking a sample down by many variables at once? What is one way to try to get around the problem, and what difficulty arises if we take this route?

5. Define the following terms: specification, moderation, distortion, spurious relationship, intervening variable, mediator.

6. What is a doubledecker? What does the width of the bars tell us?

Note

1. For those who may be wondering about random fluctuation, program A was the only one where the difference in admission rates was beyond chance expectations.

Part V
REGRESSION

Today, forms of regression are the most commonly used methods in academic quantitative sociology, and are widely used in neighbouring fields. They allow us to look at the effects of several predictors on a dependent variable with only modest increases in sample size as the number of variables expands. Regression also allows us to examine causal networks through path analysis and related methods. Because regression is so widespread, and as one writer put it, so "extraordinarily useful," it is given a full part of this volume to itself.

In Chapter 13, we consider its simplest form, bivariate regression, which involves only one dependent and one independent variable. We will look at how we choose a trend line to represent the effect of one variable on another, and at how we assess our ability to predict one from the other. The trend line itself is defined by an equation, which we will see how to interpret and to present graphically. We shall also need to see what can be done when the trend in our data cannot be represented by a straight line.

In Chapter 14, on multiple regression, we introduce additional predictors, and see how they affect the way we choose our trend line, assess our predictive success, interpret our equation, and present it graphically. In this section, we also discuss Analysis of Variance (ANOVA), an important special case.

Chapter 15, on path analysis, shows how we deal with causal networks by using a series of regression equations.

Chapter 16, on logistic regression, deals with the modifications we make when our dependent variable is a dichotomy. It concludes with an extension of logistic regression, known as multinomial regression, which allows us to deal with nominal dependent variables if they have multiple categories.

13

Bivariate Regression

Learning Outcomes

In this chapter, you will learn

- the basis on which we choose a "regression" line to represent the association between two variables;

- two measures of how well we can predict one variable from another;

- how we interpret bivariate regression results; and

- several ways to deal with associations that cannot be represented by a straight line.

Origins

The term "regression" came into statistics through Sir Frances Galton in the 1880s. In his studies of heredity, Galton had noticed that parents with extreme characteristics—for example, great height—tended to have children less extreme than themselves. He referred to this tendency as "regression to the mean." The term "regression" was transferred to a method in which we try to predict scores on one variable from scores on others. This method is now a mainstay of social data analysis.

Earlier, Galton and Pearson had used scatterplots to display correlations between parents and children, or between other sets of relatives. But scatterplots were time-consuming to produce and expensive to print. If samples were large, so cases appeared in (almost) all possible locations in the graph, trends were not clear. It was also difficult to state precisely what the difference between two scatterplots was—as Figure 13.1, based on some of Galton's original data, will illustrate. Even though the points have been jittered, without the trend line, produced by regression, only the broad tendency for height of children to rise with height of parents would have been apparent.

Figure 13.1: Heights of Offspring by Mid-heights of Parents, in Inches, Jittered, with Least Squares Line

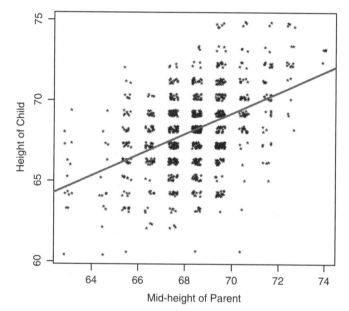

Although the broad trend could be shown, a jittered graph would still have been expensive to produce and print. A simpler alternative, readily available, was the graph of averages, which reduced the points drawn to a small number. It was possible to divide the cases into sets in terms of their scores on one variable, and then plot the average scores on the other variable for each set. To illustrate, in the graph below parents have been divided into groups on the basis of their average height, i.e., 5′ but not 5′1″, 5′1″ but not 5′2″, and so on, and we have plotted the mean heights of their children. This graph of averages in Figure 13.2 has been created from the same data used for the scatterplot above.

The graph does not include parental height at the extremes, because Galton's samples were smaller there, and hence gave unstable results. However, for the rest of the data, we have nearly a straight line. Given results like this, it seems reasonable to summarize the trend in the scatterplot with a straight line. Obviously there is variation around the line—we cannot perfectly predict the heights of children from the heights of their parents—so along with the straight line, some measure of how close the data points fall to it is needed.

Figure 13.2: Median Heights of Children by Mid-heights of Parents

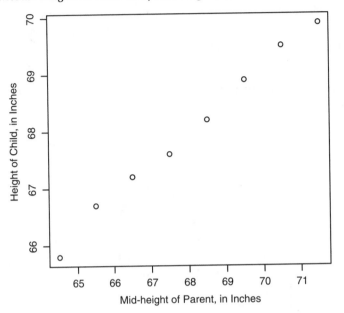

Through what we now call bivariate regression, we can define a line through a scatterplot that, in a sense, gives the best possible fit to the data, and also can measure how closely the data cluster around the line.

The Principle of Least Squares

An obvious question in summarizing a relationship with a straight line is which line to choose. If they draw a line by eye, two people may easily choose different lines. A technique giving consistent results was found in the method of least squares, borrowed from the great mathematicians Gauss and Legendre:

The principle of least squares holds that we ought to choose the line that minimizes the sum of the squared distances between scores on the dependent variable and scores predicted for them.

For any given value of X, a line will suggest a corresponding value of Y. Suppose we wish to predict a course grade from the students' entering GPAs, at a school where GPAs are scored from 0 through 13, with F− = 0 and A+ = 13. Suppose we get the equation

Course Grade = 46.667 + 3.3333*(GPA)

This can be plotted as a straight line, as in Figure 13.3.

Figure 13.3: Grade by GPA

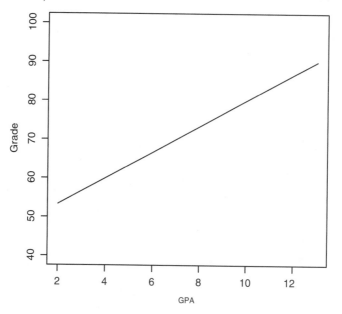

We can get the predicted grade for any GPA from the equation. Suppose GPA = 10 (for B+). Then the predicted grade will be

$$46.667 + 3.3333(10) =$$

$$46.667 + 33.333 = 80$$

Since we are thinking of incoming GPA as predicting the course grade, it is straightforward to refer to the values we get from the equations as our predicted values. We can graph these and compare them to the grades students actually get, as shown in Figure 13.4, where the predicted values are on the line and those observed are above or below it. The vertical lines show the differences between them.

We ought, by the principle of least squares, to minimize the squared differences between the predicted and the observed values of Y. More formally, we ought to minimize

$$\Sigma(\hat{y}_i - y_i)^2 \, ,$$

where $\hat{y}_i$ is the predicted value for a given case and y_i is the observed value.

The differences, denoted $(\hat{y}_i - y_i)$ are just the vertical distances between observations and the trend line.

The difference between the predicted and observed values, if our prediction is correct, will be zero. If we overestimate, $\hat{y}_i$ will exceed y_i and the difference will be positive. If we underestimate, $\hat{y}_i$ will be smaller than y_i and the difference will be negative.

Figure 13.4: Grade by GPA, Observed and Predicted Values

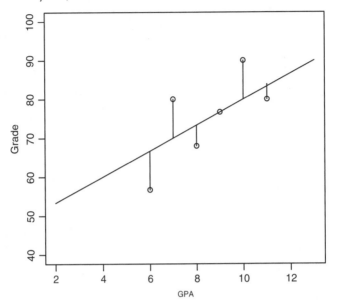

We would like to consider every case in deciding where the line should go, but we cannot simply add up our errors to see how we are doing, because positive and negative errors will cancel. To get around this, we square the differences $(\hat{y}_i - y_i)$, then add up the squared differences. The resulting sum of squared errors is the quantity we try to minimize. It is sometimes referred to this way, and sometimes as the Error Sum of Squares.

Using the principle of least squares has a number of advantages:

1. the least squares line always goes through the centroid $(\overline{x}, \overline{y})$;

2. on average, the line neither over-nor underestimates; and

3. the principle leads to two useful measures of how closely observations cluster round the line.

As a reasonable summary measure of the link between two variables, a line might be expected to pass through the centre of the data. The least squares line will always go through the point at which both x and y take on their mean value, the centroid, which is denoted $(\overline{x}, \overline{y})$.

Again, it seems reasonable to expect a measure used to summarize a relationship neither to over-nor to underestimate the values of the dependent variable, on average. The least squares line will over-or underestimate for individual cases, but these errors will cancel out perfectly. In algebraic notation,

$$\Sigma (\hat{y}_i - y_i) = 0$$

The Logic of the SEE

The two measures of how well the data cluster around the line are known as the standard error of estimate (SEE) and r^2. Each will require some explanation.

To show how they are obtained, we will begin from the sum of squared errors. As a measure of how well we can predict the values of Y, it suffers from the drawback that larger samples tend to produce greater sums, so we cannot straightforwardly compare the sum from one study to the next unless the number of cases is the same.

To get a comparable figure, we can average the squared errors in each sample, dividing by N,[1] the number of observations, to get a mean squared error. The result,

$$\Sigma\,(\hat{y}_i - y_i)^2\,/\,N\,,$$

is sometimes called just that, or sometimes the Error Variance.

This term comes from the similarity between this formula and the formula for the Variance, which for y is just

$$\Sigma\,(y_i - \bar{y})^2\,/\,N$$

The difference is that in the Error Variance, instead of comparing y_i with $\bar{y}$, we compare it with $\hat{y}_i$.

But there is still a problem to work around. As we have seen when discussing the standard deviation, when we square a quantity, we also square its units. Here, our errors of prediction will be in squared marks, which make no sense.

To get around the problem with units, we take the square root of the Error Variance. In doing so, we "desquare" the units as well as the number associated with them, and we get a measure expressed in the natural units of y. This measure is known as the standard error of estimate (SEE), often denoted $s_{y.x}$.

$$SEE(s_{y.x}) = [\Sigma\,(\hat{y}_i - y_i)^2\,/\,N]^{.5}$$

In fact, the SEE is the standard deviation of the errors of prediction. Above, we saw that the SD of x is given by

$$[\Sigma\,(x_i - \bar{x})^2\,/\,N]^{.5}$$

To get the SD of the errors, we substitute $(\hat{y}_i - y_i)$ for x_i. Since the mean of the errors is 0, we insert 0 for $\bar{x}$, and obtain

$$[\Sigma\,((\hat{y}_i - y_i) - 0)^2\,/\,N]^{.5} =$$

$$[\Sigma\,(\hat{y}_i - y_i)^2\,/\,N]^{.5}\,,\ \textbf{which is the SEE}$$

Above, we saw that if a distribution is normal, 95% of the observations lie within 1.96 standard deviations of the mean. If our errors are normally distributed, 95% of them will lie within 1.96 SDs, here called standard errors, of their mean, which is zero. That is to say, 95% will be no greater than ± 1.96 standard errors. Putting it the other way round, only 5% will be > ± 1.96 standard errors.

As shown above, if a distribution is unimodal and strictly continuous, no more than 11.1% of observations will lie more than two SDs from the mean. Thus the distribution of errors can depart considerably from normality while keeping the vast majority within the range of ± 2 standard errors. In practice, we regularly find, when working with social survey data, that over 90% of our errors lie within this range.

For example, consider Figures 13.5 and 13.6. The first shows the residuals obtained by predicting household income at one point in time from household income three years earlier. The distribution has asymmetric tails, and is a bit leptokurtic, but 94.6% of the residuals lie within 2 SEs of the mean.

Figure 13.5: Histogram of Income Residuals

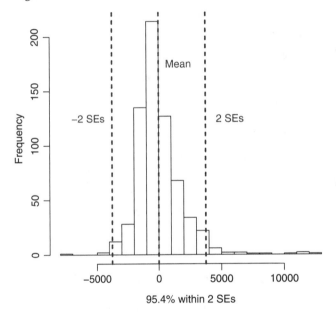

Figure 13.6 shows the residuals from a prediction of depression using family functioning as the independent variable.

These residuals are right skewed, sufficiently that none of them lie more than 2 SEs to the left of the mean. Those to the right make up just 5.3% of the lot, so 94.7% are smaller than ± 2 SEs.

The Logic of r^2

Another widely used measure of how closely the data hug the line is known as r^2. It is the square of Pearson's r, and a PRE measure. You will recall that to create a PRE measure we take

$$((Error1) - (Error2)) / (Error1)$$

This gives us the proportion by which we can reduce our errors in predicting a DV if we know how the DV and the IV are linked for individual cases.

Figure 13.6: Histogram of Depression Residuals

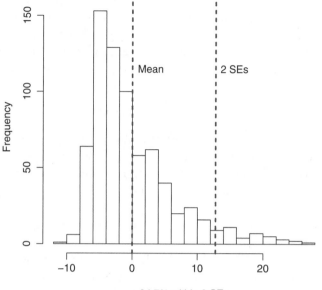

94.7% within 2 SEs

We have seen above that the Error Sum of Squares and the Error Variance are based on squared differences between our predicted and observed values of y. To combine with one of these in a PRE measure, we need a measure of squared differences between the predicted and the observed values of y when we do not know how x and y are linked.

If we do not, we can still calculate the mean of the DV and use it in estimating the value of individual cases. In fact, knowing nothing else, we will do better by estimating each case to take on the value of the mean than by any other strategy. (Better in the sense that the sum of our squared errors will be least—a proof of this point was presented above, on pages 23–24.) When we do this, our error for each case can be written

$$(y_i - \bar{y})$$

If we want an overall measure of how well we are doing, we will somehow have to add up our errors across cases. To avoid having positive and negative errors cancel, we will first square the term. The resulting quantity can be written as

$$\Sigma (y_i - \bar{y})^2$$

This quantity is referred to as the Total Sum of Squares. To get a quantity that doesn't depend on sheer sample size, we can divide by N and get

$$\Sigma (y_i - \bar{y})^2 / N$$

This is our mean squared error of prediction. It is also the Variance of y!

We now have measures based on our squared errors with and without knowledge of the link between x and y. The Total Sum of Squares and the Variance of y apply when we take no IV into account. The Error Sum of Squares and the Error Variance apply when we take an IV into account. In setting up a PRE measure we can use either the Variance and Error Variance or the Total Sum of Squares and the Error Sum of Squares. (The two sets of figures differ only in that to get the Variances we divide the Sums of Squares by a constant.) Here we shall use the Variances.

$$r^2 = (Var(y) - Error\ Var(y)) / Var(y)$$

The interpretation of r^2 follows directly from its PRE quality. It gives us the proportion by which we can reduce our squared errors in predicting y if we know how it is linked to x.

You will often see this phrased another way. It is often said that r^2 tells how much of the Variance in y we can explain from x. If to explain something means to be able to predict it, this expression makes sense. If $r^2 = .50$, then we have reduced our squared errors of prediction by 50%, and we have, by the same token, improved our ability to explain y by 50%.

We may also consider the extent to which the values of x determine those of y. A measure of our ability to predict the value of y from x is a measure of its determination by x. Thus r^2 is sometimes referred to as the coefficient of determination.

Relative Usefulness of the SEE and r^2

The two measures of how well we are doing have contrasting strengths. The SEE gives us a measure of error in the natural units of the DV. On the other hand, r^2 has a very helpful PRE interpretation. But r^2 is vulnerable to another problem. It will automatically rise as the standard deviations of the two variables rise, so that it cannot readily be compared across samples without a close look at the SDs. The graphs in Figure 13.7 illustrate: the two b's are very close, as are the SEEs. However, the SDs in the second graph are much larger than those in the first. R^2 rises from .67 to .90 because of this.

Figure 13.7: Changes in r^2 Produced by Differing SDs

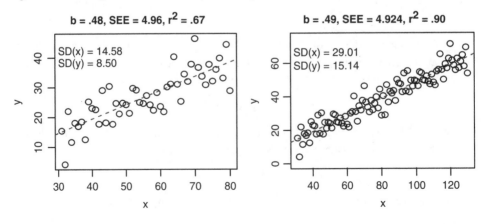

The SEE, though, is not directly affected by the SDs and thus may be readily compared. It is for this reason that some consider it the stronger measure. It is not, though, the most widely used, because the PRE interpretation of r^2 is very attractive. Perhaps the sensible thing is to use them both, to take advantage of their complementary interpretations, whenever we are not comparing across samples.

The Form of the Equation

From your earlier geometry, recall that an equation of the form

$$y = a + bx$$

defines a straight line. Y is to be plotted on the vertical axis (or y-axis) and x on the horizontal axis (or x-axis) of the graph.

To see what the first term on the right (a) does in the equation, notice what happens if x = 0. Then

$$y = a + b(0) = a$$

Thus the first term gives us the value of y when x = 0, the value of y when the line crosses the y-axis, or in other terms, the value of the "intercept." (The latter term just refers to the point at which the line crosses the y-axis.)

If we know one point through which a line passes, and the direction in which it is going, we can determine all other points on the line. We know from the intercept one point through which the line passes. Its direction is given by b, which gives us the slope of the line. Its most general interpretation is that it gives us the number of units of change in y for one unit of change in x.

Suppose that we want to plot income by years of education and we have an equation of the form

$$y = a + bx ,$$

which, when solved, gives

$$\text{Income} = 10{,}000 + 3{,}500*(\text{Years of Education})$$

If Years of Education takes on the value of 10, the equation yields

$$\text{Income} = 10{,}000 + 3{,}500(10) = 10{,}000 + 35{,}000 = 45{,}000$$

Suppose now that Years of Education becomes 11. Then

$$\text{Income} = 10{,}000 + 3{,}500(11) = 10{,}000 + 38{,}500 = 48{,}500$$

The change in predicted incomes when Years of Education is 11 rather than 10 is (48,500 – 45,000) = 3,500, the value of b.

Suppose now that Years of Education becomes 12. Then we have

$$\text{Income} = 10{,}000 + 3{,}500(12) = 10{,}000 + 42{,}000 = 52{,}000$$

The difference in predicted incomes is now $(52{,}000 - 48{,}500) = 3{,}500$, the value of b once again.

You may see another interpretation, which is that b gives us the rise over the run, where the rise is the change in y from one point on the line to another, and the run is the change in x between the same two points. A little algebra will show that this interpretation is equivalent to the first.

Interpreting b

We have been speaking of the meaning of the coefficients in a quite general equation for a straight line. For regression, we will change the notation to

$$\hat{y} = a + bx$$

We switch y to ŷ because the line gives us the predicted value of y for each value of x and the caret, or "hat" symbol, is used for predicted values.

A retains its interpretation as the value of the intercept. B can simply be interpreted as the number of units' change in the predicted value of y for one unit of change in x. Since some of these predicted values will be high and others low, but over- and underestimates will cancel out, we can say that

b gives us the average number of units of change in y for a unit of change in x.

B is referred to in several ways: as the slope or the slope coefficient, as the regression coefficient or the regression weight, or as the b-coefficient. The terms are strictly synonymous.

You will also see references to betas. These too are regression coefficients but of a particular kind. Sometimes we work with variables which have been standardized, or z-scores. (Recall that these are created by subtracting the mean from the score of each observation, then dividing the remainder by the standard deviation.) These always have a mean of zero and a standard deviation of one. When we have standardized both variables, the b's are ordinarily referred to in the social sciences as betas (βs).[2] When the variables have not been standardized, the b's can be called "metric" coefficients. In the sociological journals today, metric coefficients are far more common, but betas must be understood, particularly if we wish to read in neighbouring disciplines, particularly psychology.

Betas give us the average number of SDs' change in Y for one SD of change in X. Suppose we have standardized income and standardized years of education and we solve

$$Z_{\text{Income}} = A + \beta^*(Z_{\text{Years of Education}}) \, ,$$

Obtaining

$$Z_{\text{Income}} = 0 + .35(Z_{\text{Years of Education}})$$

Then we know that, on average, for one SDs' increase in years of education, we get .35 SDs of increase in income.

The intercept is 0, as is always the case when the variables are standardized. The means of standardized variables are always 0, and the least squares line always goes through the centroid $(\bar{x}, \bar{y})$. Thus it passes through the point (0,0). This is where the line cuts the y-axis (since x = 0), so A = 0.

It was pointed out above that Pearson's r has the same interpretation we have now seen for beta. Their equivalence can be seen from their respective formulae.

$$r = Cov(x,y) / [SD(x) \, SD(y)] \text{ , as we have seen above}$$

$$b = Cov(x,y) / Var(x)$$

Their numerators are the same. B will equal r, then, whenever the denominators are the same, i.e., when SD(x)SD(y) = Var(x). Since Var(x) is just the square of SD(x), for standardized variables it will be $1 \times 1 = 1$. Because of the standardization, SD(x)SD(y) will also be $1 \times 1 = 1$. Thus, in this special case, r = b.

You may also, if not commonly, encounter situations in which the single IV is dichotomous, as in Canadian official language (anglophone, francophone) or sex (male, female). In this situation the interpretation is simple, provided the two groups have been assigned scores one unit apart: b simply gives the average difference between the two groups. In fact, this interpretation is just a special case of the more general one set out above. Suppose we have coded sex as 0 = M, 1 = F, and obtained the equation:

$$\text{Income} = 50,000 - 10,000*(\text{Sex})$$

If a respondent is male, the equation becomes

$$\text{Income} = 50,000 - 10,000(0) = 50,000 - 0 = 50,000$$

For a female, it becomes

$$\text{Income} = 50,000 - 10,000(1) = 50,000 - 10,000 = 40,000$$

The difference, 10,000, is just the value of b. So, as usual, the coefficient gives us the amount of change we get, on average, in the DV for one unit of change on the IV. It is just that, the way we have coded it, one unit on the IV represents a shift from one group to the other. When dealing with a dichotomous IV, we typically code it with the two groups one unit apart, so we can interpret b as the average difference between them.

The only thing to watch out for is the sign of the coefficient, which gives us the direction in which the DV changes as we move from the group coded 0 to the group coded 1. In this example, the sign is negative, so it means that income declines as we go from those coded 0 (male) to those coded 1 (female). The catch is that the coding of such variables is quite arbitrary. There is no logical reason why we shouldn't have coded this one the other way around, with F = 0 and M = 1. If we had, we would have obtained the equation

$$\text{Income} = 40000 + 10,000*(\text{Sex}) \text{ ,}$$

so for males, we could have calculated

$$\textbf{Income} = \textbf{40,000} + \textbf{10,000(1)} = \textbf{40,000} + \textbf{10,000} = \textbf{50,000}$$

and for females, we would have calculated

$$\textbf{Income} = \textbf{40000} + \textbf{10,000(0)} = \textbf{40,000} + \textbf{0} = \textbf{40,000}$$

We would have gotten exactly the same predicted values. It is just that the sign of the coefficient would have changed, together with the value of the intercept. There is a general rule: *the predicted values do not change if we reverse the coding of a variable, but the sign of the coefficient does.* Thus, to interpret the coefficient we have to look at the coding of the IV.

This point holds for predictor variables which are not dichotomous, and for dependent variables. Suppose, for example, we want to predict how satisfied people are with their jobs from their incomes. The satisfaction measure could be coded so that those with high scores are highly satisfied or highly dissatisfied. Unless we know how it has been done we will be unable to interpret the sign on b.

Graphic Display for Bivariate Regression

To show people unfamiliar with regression how two variables are related, we can put the regression line on a scatterplot, as in Figure 13.8. Here we show scores on a scale for depression by scores on a scale of perceived social support. Individual scores have been jittered so we can see where there are concentrations of cases. Unsurprisingly, as perceived support rises, scores on depression decline.

Figure 13.8: Depression by Social Support, b = −.657

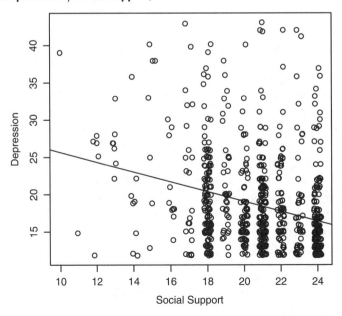

Non-linear Trends

Thus far we have looked at variables whose relationships can be represented by straight lines. However, there is no reason why we should consistently be able to represent trends with straight lines. We very often can, perhaps surprisingly often can, but other possibilities must be dealt with. If we encounter a non-linear trend, we can often transform a variable to create a linear relationship. Common transformations will be illustrated here.

Truncation

One involves a trend line which goes flat, either on the left or on the right. In Figure 13.9, the link between income and education goes flat on the left.

Figure 13.9: Income by Years of Education with a Truncated Effect (Hypothetical Data)18

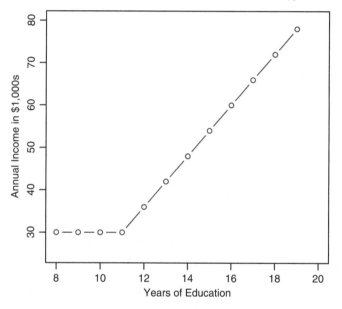

In these hypothetical data, income rises strongly for those who have gone past Grade 11, but income is not affected by a rise from Grade 8 to Grade 11.

In such cases, we often truncate the independent variable at the point at which the trend flattens out. Here a data analyst might well recode 8, 9, and 10 to equal 11, then refer to the new category as "up to 11." Since the categories below 11 would then disappear, a straight line would appear on a revised graph. The equation for the line, which represents income in 1,000s, would become

Annual Income = −30 + 6*(Years of Education)

The accompanying text would explain that education had been truncated at 11, and that the equation should be interpreted as meaning that from 11 years onward income rises by 6,000 for each additional year of education.

For an example with real data, Draisey (2013) found that, unlike what we would expect in Canada, in Cambodia, Body Mass Index rose up to the age of 40, but not thereafter. Because the sample was so large, including dots for every case tends to obscure the trend line, so only the line is shown.

Figure 13.10: BMI by Age

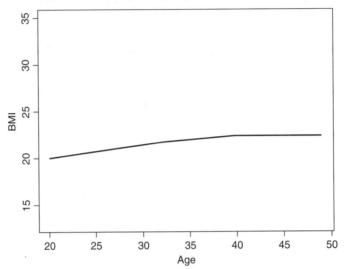

Here the truncation took place at 40. The highest remaining value was relabelled "40+." The equation now indicates how much BMI rises per year of age up to 40.

The modified equation is

$$\text{BMI} = 18.537 + .086 \, (\text{Age}) \, ,$$

so for someone 40 or over, it predicts

$$\text{BMI} = 18.537 + .086 \, (40) = 18.537 + 3.440 = 21.977$$

Exponential Curves

In another fairly common situation, as the IV rises the DV rises at either an accelerating or a decelerating rate. Figure 13.11 provides an example.

In these hypothetical data, income does not rise by a constant number of dollars. Rather, it rises at a constant percentage rate, of 9% per year. So, for those with 9 years rather than 8, income rises from 31,000 to 32,790. A 9% increase applied to that figure yields a larger rise than the first one because the base on which the 9% increase is calculated is now greater than the initial 31,000. Because the base for the 9% rise is continually increasing we have an accelerating curve, technically called an "exponential" curve.

Figure 13.11: Annual Income by Years of Education, With an Exponential Effect (Hypothetical Data)

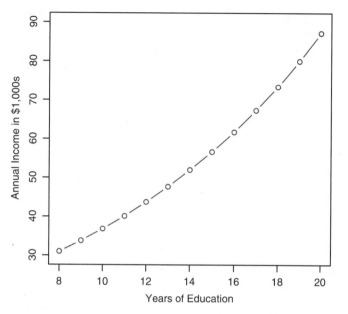

Recalling that the opposite of "exponentiating" is "taking the log," a way to tame an accelerating curve is to take the log of the accelerating variable. Figure 13.12 presents the result of logging income.

Figure 13.12: ln(Income) by Years of Education (Hypothetical Data)

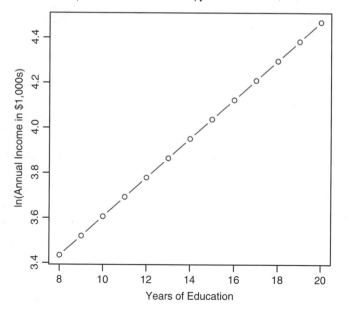

(If you would like a refresher on logs, notes about them are found in Appendix B.)

With real data we do not expect such perfect results, but as will be seen in a moment, logging an accelerating variable is often very helpful. Apart from straightening out the line, it often gives us a useful interpretation for b.

As long as b < .20, b gives us (approximately) the percentage increase we obtain in y for a unit of change in x.

In this case, if we regress In(income) on years of education,

$$b = .086 \, ,$$

suggesting that we have an increase of about 9% in y for a unit of change in x, a result that is not far from the true figure.

For another example, let us consider Canadian population growth. Figure 13.13 displays the data.

Figure 13.13: Canadian Population in Millions, 1851–2011

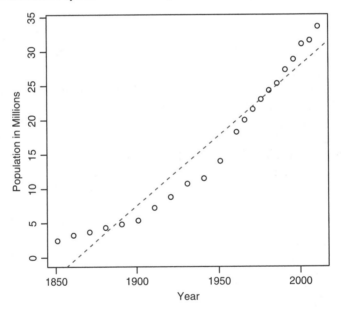

Plainly, we have an accelerating curve. Taking the log of population yields an almost straight line, as shown in Figure 13.14.

Further, we have an interpretable b. For a single year, b = .01678. We do not expect major change in a single year, but a decade may make a considerable difference. For a lapse of 10 years, we multiply b by 10, obtaining .1678. This figure suggests about a 17% increase in the population over a period of that length.

More Than One Slope

Sometimes the effect of x changes somewhere along its range. Perhaps, for example, the value of an additional year of education is greater for those who have more than the median level than for those with less. If so, a graph like Figure 13.15 ought to result.

Figure 13.14: ln(Canadian Population in Millions) 1851–2011

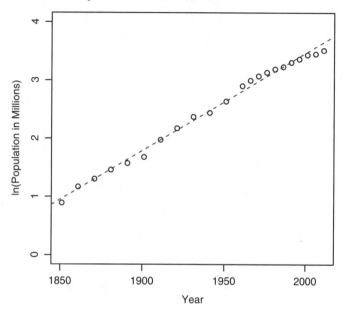

Figure 13.15: Annual Income by Years of Education, with a Changing Slope (Hypothetical Data)

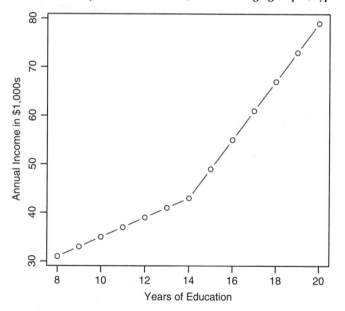

Here income is represented in 1,000s of dollars. Up to 14 years of education, the graph shows a rise of 2,000 for each additional year. From 14 on, the rise is 6,000.

When a change of slope occurs, we need a way to get a second b.

This is only possible if we have a second predictor. We obtain one by creating a "spline." There is more than one way to do this, but we usually create a variable whose b will give us the change in slopes at the "knot," the point at which the slope alters.

Here the knot lies at 14. The spline will take the value of 0 up to the knot. Above the knot it will rise by one unit for every rise of one unit in the original variable. The values of the original and the spline will be as follows:

for the original: 15   16   17   18   19   20

for the spline: 1   2   3   4   5   6

(We get these results by subtracting the value of the knot, 14, from the original values.)

Because we now have two variables, the original and the spline, we will get two b's. For this example, we will obtain the equation

$$\text{Income} = 15 + 2*(\text{Years of Education}) + 4*(\text{Education Spline})$$

The first b (2) gives us the change in income for a year of additional education up to the knot. The second b (4) gives us the change in the slope. Since the initial slope is 2 and the change is 4, the slope beyond the knot is 2 + 4 = 6.

A further example uses data from a Canadian post-secondary program, in which grades in the student's last year of high school math were used to predict grades in an initial statistics course. The results are shown in Figure 13.16.

Figure 13.16: Grades in Initial Statistics Course by High School Math Grades

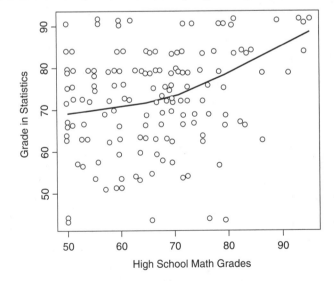

Plainly the trend up to a high school grade in the low 70s differs from the trend for those with grades in the high 70s and above. A smooth sometimes rounds off a corner in a way that makes it difficult to choose the best knot, but here it will be placed at 74.

The resulting equation is

Statistics Grade = 64.045 + .144*(HS Grade) + .643*(HS Spline)

Up to the knot at 74, statistics grades rise, on average, by .144 for each additional high school mark. At the knot, the slope changes by .643, so the slope beyond 74 is .144 + .643 = .787. Up to 74, high school grades are a quite weak predictor. Beyond that point they are much stronger.

Handling Nominal Polytomies

We often wish to use a polytomous nominal variable as a predictor. For example, if we wanted to predict earned income, we might want to see how much difference there was between those born in Canada and those born in other places. Of course, we might just want to compare immigrants and the native-born, but we might want to break down place of birth much more finely. If we had a national sample of, say, 10,000, it might well sustain the use of 10 to 15 categories. Suppose we wanted to distinguish among the following locations.

1 - Canada

2 - US

3 - Caribbean

4 - elsewhere in the Americas

5 - Europe

6 - Southeast Asia

7 - China

8 - elsewhere in Asia

9 - elsewhere

We could not use these categories as they stand, because the scores for the categories are arbitrary. If asked what a one unit on this variable meant, we would be at a loss, and so we could not interpret the b for it. Fortunately, we have a way out.

In cases like this, we can represent each category, save one, by a "dummy variable." For example, for category 3 "Caribbean," we could create a variable with two values:

0 - not born in the Caribbean area

1 - born in the Caribbean

Similarly, for category 5, we could create a variable coded

0 - not born in Europe

1 - born in Europe

And so on, for all the categories save one.

We leave out one for two reasons. First, if there are g categories, we need only g − 1 variables to distinguish among them. Suppose that we do not create a variable specifically representing Canada. We can still tell those born in Canada, because they are the ones with a code of 0 for each of the variables we have created.

The second reason is more technical: if we use a dummy for each category the processes that define our equation will end up trying to divide by zero. To avoid this, statistical packages either toss out a variable or give us an error message and stop.

When we use two or more dummy variables (or dummies, for short), we refer to the category without its own as the reference category. The choice of reference category is important because the b's for the dummies we create give us the difference between the categories they represent and this reference category. So, for example, if we use "Canada" as the reference, then the coefficient for "Caribbean" will give us the mean difference between those born in the Caribbean and those born in Canada, and the coefficient for "Europe" will give us the mean difference between those born in Europe and those born in Canada.

To see this, we need only look at an equation. Suppose that we try to predict life satisfaction, measured on a scale from 0 to 10, from country of birth; that we use Canada as the reference category; and that we obtain

$$\text{LifeSat} = 7.50 + .20*(\text{Europe}) − .25*(\text{SE Asia}) \ldots + .15*(\text{Elsewhere})$$

We can calculate the predicted score for any category from the equation. Each group starts from 7.50, the value of the intercept, then gets the value of whichever coefficient applies. Let us see how this works out for three different places of birth.

$$\text{for Canada, LifeSat} = 7.50 + .20(0) − .25(0) + .15(0) = 7.50$$

$$\text{for Europe, LifeSat} = 7.50 + .20(1) − .25(0) + .15(0) = 7.70$$

$$\text{for SE Asia, LifeSat} = 7.50 + .20(0) − .25(1) + .15(0) = 7.25$$

Since no dummy singles out Canada, the final predicted score equals the intercept. For Europe and SE Asia, the departures from the intercept are equal to the coefficients for the dummies representing them. Thus the difference between Europe and the reference category, or between SE Asia and the reference category, has been given by the b for its dummy.

Quadratic Trends

The final transformation considered here is the quadratic. It is awkward to provide a plausible hypothetical example based on education and income, so we will look at household income and age, as shown in Figure 13.17.

Figure 13.17: Household Income by Age, with a Quadratic Effect (Hypothetical Data)

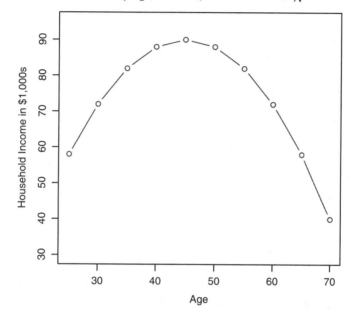

Here, income rises from the mid-20s to a peak in the mid-40s, then declines.

A quadratic (squared) term produces a single broad curve, such as we observe here.

Quadratics yield equations that are typically difficult to interpret verbally. Here, with income in 1,000s, the equation is

$$\text{Household Income} = -72.00 + 7.20*(\text{Age}) - .08*(\text{Age}^2)$$

Clearly, the interpretation of the coefficients is more difficult here than for the other transformations we have seen. Typically authors say "the relationship is quadratic," or perhaps they indicate the direction of the bend by say that it is "quadratic upward" (or downward if that is the case). To show clearly how a quadratic term affects the DV, it is typically necessary to use a graph.

An alternative, to be considered if we need a straightforward verbal interpretation, is to use a series of dummy variables. To see how that can work out, let us look at the link between age and household income in a real data set (the Canadian Election Survey of 2008).

In Figure 13.18 means are presented for each of seven age categories: up to 24, 25–34, 35–44, 45–54, 55–64, 65–74 and 75+.

Figure 13.18: Household Income by Age Category

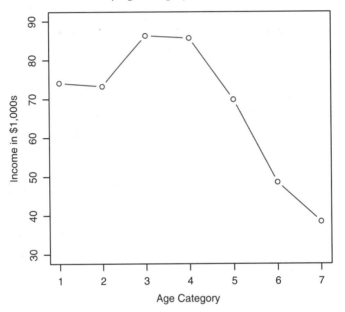

Apart from the first category, the graph resembles a quadratic curve, but the exception is a great enough problem that we will prefer to use a set of dummy variables for the age categories. As we have seen earlier, when we convert a variable to a set of dummies we use one less dummy than there are categories.

We also have to choose a reference category. Here we shall use the first category, up to 24. The resulting equation is

$$\text{Household Income} = 74.12 - .86*(25\text{–}34) + 12.18*(35\text{–}44) + 11.57*(45\text{–}54) \\ - 4.28*(55\text{–}64) - 25.59*(65\text{–}74) - 35.66*(75+)$$

Since the reference category (18–24) has no dummy, those in this category are predicted to have a mean income of 74.12 thousand.

The dummies for other categories give us the mean differences between them and the reference category. Those 25–34 do not differ appreciably from those under 25. The next category (35–44) stands 12.18 thousand higher than the reference category, and those 45–54 are in almost the same position. Then we see the beginning of a decline. Those 55–64 are lower than the reference category by 4.28, those 65–74 are below by 25.59 and those 75 and over are below by 35.66.

Because we have been using dummy variables, we have been able to give a verbal comparison of the reference category with each of the others. This way of discussing differences is apt to be clearer to many people than what could have been obtained from a quadratic transformation.

Summary

In bivariate regression we try to sum up the link between two variables with a trend line, chosen to minimize the sum of squared errors in predicting y. This "least squares line" can be represented by an equation including an intercept, which gives the value of y when x = 0, and a slope, which tells us how much y changes, on average, for a one-unit change in x. Choosing the least squares line leads to two by-products: r^2, which tells us how much of the Variance in y is accounted for by x, and the standard error of estimate, which is the standard deviation of the residuals.

   When we cannot represent the link between variables with a straight line, we have several options:

* we can truncate a variable (when the trend line flattens at some point);

* we can take the log of a variable (to deal with an accelerating curve);

* we can create a spline (so we can represent two different slopes);

* we can create dummy variables (to represent different categories of cases); and

* we can create a quadratic term (when there is a single broad curve of quadratic form).

These same strategies of transformation may be used in multiple regression, which is the most used analytic strategy in quantitative sociology today. We will turn to it in Chapter 14.

Review Questions on Bivariate Regression

1.   Suppose we try to predict marks on the final exam (Final Exam) from hours of studying in the week before it (Study Hours). Suppose we obtain the equation

   Final Exam = 30 + 2.2*(Study Hours)

   What does the 30 tell us? The 2.2?

2.   In the general formula y = a + bx, what are "a" and "b" called? (Give two distinct names for "b").

3.   Suppose we try to predict final standing in a course like this (Final Mark) from scores on a test of math anxiety (Math Anxiety) on which a score of 0 means no measurable anxiety and a score of 100 represents the opposite extreme. Suppose we obtain the equation

   Final Mark = 60 + .03*(Math Anxiety)

   What does the 60 tell us? The .03?

4.   Briefly explain the principle (least squares) through which a regression line is typically chosen.

5. What are two helpful "by-products" of our usual method of choosing a bivariate regression line?

6. Is the standard error equal to the Variance of our errors of prediction? If not, what does it equal?

7. If our errors of prediction are normally distributed, within what range will 95% of them lie?

8. Suppose that for the equation in question 3, $r^2 = .05$. How do we interpret this?

9. Show why Pearson's r is equal to the regression coefficient for standardized variables.

10. What is the "generic" interpretation for b? What does it become when the variables are standardized?

11. If a predictor is dichotomous (e.g., male/female), what does b tell us?

12. What is truncation? When might we apply it in regression?

13. What might we do to deal with an accelerating curve?

14. How do we typically interpret the coefficient for a spline? Why are these created?

15. When does b give us an estimate of how much % change we get in y for a unit of change in x?

16. What is an alternative to a quadratic curve? What is an advantage to the alternative?

Notes

1. Working with sample data, we typically divide by N − 1, but the difference, except for small samples, is trivial and will be neglected here.

2. Betas typically mean something else in theoretical statistics, but in social science today the b's for standardized variables are usually referred to this way. You may occasionally see b*s.

Multiple Regression

Learning Objectives

In this chapter, you will learn

- how we estimate the effects of several predictors on a dependent variable at once;

- similarities and differences between bivariate and multiple regression;

- what may happen if we do not use the right set of predictors;

- how we interpret a table of regression results;

- what we do if the effects of one predictor depend on the value of another;

- how to understand the special case of Analysis of Variance; and

- how to set up a regression table.

Why Multiple Regression (MR)?

Bivariate regression (BR) has important advantages:

1. it provides a measure of association between two interval or ratio variables;

2. if the association is causal, b provides a measure of causal impact; and

3. it provides measures of how well one variable predicts the other, in the SEE and r-squared.

But typically it must be extended, because

1. we need more than one IV to predict the DV well;

2. we need to know how much impact more than one IV has; or

3. we want a clean reading on the effect of an IV, holding other potential predictors constant.

The need to hold other predictors constant may be clearer with an example. Suppose we continue with predicting depression. As independent variables, we have both family functioning and sense of social support, which are correlated ($r = .532$). If we estimate

$$\text{Depression} = a + b(\textbf{Family Functioning}),$$

family functioning, because it is correlated with social support, will get credit for part of the influence of the latter on levels of depression. Similarly, the social support measure, if used alone, will get credit for part of the effect of family functioning.

In general, if two predictors are correlated, and each is correlated with the dependent variable, each, if used alone, will get credit for part of the impact of the other. By using multiple regression, we can get the effect of each, with the other controlled, as illustrated in Figures 14.1 and 14.2.

Figure 14.1: Depression by Family Functioning, with and without Social Support Controlled

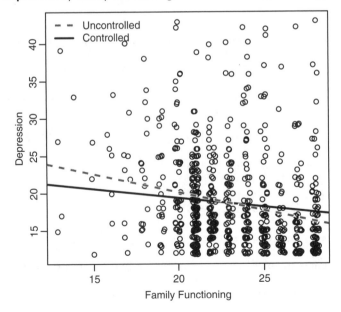

Figure 14.2: Depression by Social Support, with and without Family Functioning Controlled

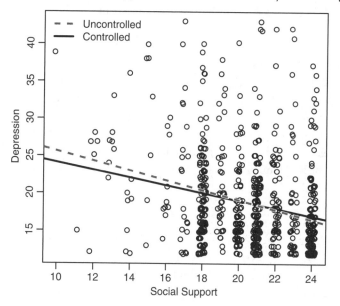

More extreme examples are easily found. One Canadian graduate student who was looking into fear of crime found in a sample of undergraduates that fear of criminal victimization was strongly related to height. When he controlled for the sex of the student, the relationship with height vanished.

Multiple regression (MR) allows us to add more than one additional predictor, and gives us the impact of each, unaffected by its links to the others. To understand MR, one must know how their individual impact is obtained.

Obtaining b's in Multiple Regression

A central difference between bivariate and multiple regression lies in how the b's are estimated. When predictors are correlated, we need to know which should get credit for how much of their common predictive capacity. The solution is obtained through matrix algebra, but the results come out as if we had done our work in stages. It is as if we had initially obtained a measure of each IV which was uncorrelated with each other IV, and then used this purified measure to predict the DV.

To get a measure of a given IV which is unrelated to the other IVs, we can regress each IV on the others. Suppose we have two IVs, x1 and x2. Then initially we can solve the equation

$$x1 = a + b(x2)$$

When we predict x1 from x2 there will be errors in our predictions (unless they are perfectly correlated, in which case we shouldn't be trying to use them both). These errors can be denoted

$$(\hat{x}1_i - x1_i)$$

They may be referred to as the *residuals* obtained by regressing x1 on x2.

As long as x1 and x2 are in a linear relationship, the residuals for x1 will be independent of x2, in the sense that they cannot be predicted from x2. They represent the part of the variation in x1 that is independent of x2. If we use them to predict the DV, they will thus give us a reading on how x1 affects y, with x2 out of the picture. (There are other ways to say this: with x2 held constant, with x2 controlled, independent of x2, and so forth.)

We can do the same thing with x2 by regressing it on x1. We obtain residuals, which we can denote as

$$(\hat{x}2_i - x2_i)$$

Again assuming that the x-variables are in a linear relationship, these residuals are independent of x1. Using these to predict Y, we get a measure of the effect of x2 on y with x1 controlled (or held constant, or taken out of the picture).

What happens in MR, in effect, is that we predict each IV from the others, obtain residuals, and use them to get the impact of an IV with the others held constant. We don't have to do this ourselves—it happens through matrix manipulations carried out by software when we ask it to carry out a regression—but behind the scenes, this is what is happening.

Given the estimated impacts of IVs, we can write equations of the form

$$\hat{y} = a + b_1(x_1) + b_2(x_2) + \ldots + b_N(x_N) \text{ , where}$$

$\hat{y}$ is the predicted value of the DV,

$x_1, x_2 \ldots x_N$ are the N IVs, and

$b_1, b_2 \ldots b_N$ are the N coefficients for the IVs.

Using Least Squares in Bivariate and Multiple Regression

Many things remain as they were in bivariate regression, and many others are only slightly modified. To begin with, the equation is defined, as before, by the principle of least squares. That is, we minimize the quantity

$$\Sigma(\hat{y}_i - y_i)^2 \text{ ,}$$

the sum of squared errors (or the Error Sum of Squares).

As before, an equation chosen in this way has three desirable properties:

1. it passes through the centroid, the point at which the variables take on their mean values (although now it passes through the means of more than two variables at once.)

2. it neither over- nor underestimates, on average. More formally,

$$\Sigma(\hat{y}_i - y_i) = 0$$

3. it will lead to two different measures of how well we are doing in predicting the DV.

As before, to minimize the sum of squared errors will be to minimize the Error Variance, which is just the mean of the squared errors.

As you will recall, while this has a straightforward interpretation as the mean of the squared errors, it comes in squared units, which usually do not make sense. So we "desquare" the units by taking the square root, and arrive at the SEE. In multiple regression, it may be written $s_{y.1..k}$, where the k in the subscript is the number of predictors being used.

As before, this is the standard deviation of the errors of prediction. As before, if the errors are normally distributed, 95% will lie in the range ± 1.96 standard errors. Even if their distribution is only unimodal and continuous, at least 88.9% will lie in the range ± 2 standard errors.[1]

In practice, as in bivariate regression, we typically find that 90% or more lie in the range that applies for normally distributed errors. Figure 14.3 displays the residuals from a prediction of household income using several independent variables, including average income on two previous occasions; level of education of the respondent; number of adults with full-time jobs; and number with managerial, professional, or technical jobs. Although the residuals are leptokurtic and a little skewed, 94.7% lie within 2 SEs.

Figure 14.3: Histogram of Income Residuals

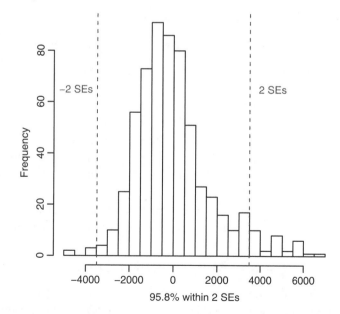

Just as the principle of least squares leads to the SEE, it also leads to a PRE measure of how well we can explain the DV. It is called R^2 to distinguish it from r^2, since it is not the square of Pearson's r between two variables. R^2 does have the same PRE basis though. It represents

$$[Var(y) - Error\ Var(y)] / (Var(y))$$

The only difference is that the Error Variance here is based on more than one predictor.[2]

The relative strengths and weaknesses of the SEE and of R^2 are those we have seen earlier. The SEE gives us a measure of how closely points are clustered around the regression line which is expressed in the natural units of the DV. R^2 has a convenient PRE interpretation. But unfortunately R^2 tends to rise with the SDs of the variables involved, so that unless we examine them, it is difficult to use in comparisons across samples. The SEE is not directly affected by the SDs and hence can be readily compared across samples. Since the two measures are complementary, it is wise to use them both where they are appropriate.

Interpreting b's in Multiple Regression

Obtaining b's as we do requires a change in interpretation. Now the standard interpretation must be that

b gives us the average number of units' change in Y for one unit of change in an IV, <u>holding other IVs constant</u>.

Suppose we solve an equation with one non-dichotomous predictor and one dichotomy,

Income = a + b$_1$ (Years of Education) + b$_2$ (Sex) , and we obtain

Income = 30,000 + 1,500*(Years of Education) − 10,000*(Sex)

where the first IV is coded in years of education completed and the second is coded: 0 = M; 1 = F.

Then the coefficient of 1,500 for Years of Education implies that for a one-year increase in education we get, on average, a $1,500 increase in income, independent of sex. The coefficient of −10,000 for sex indicates that, on average, women get $10,000 less than men, with years of education controlled. Here as in BR, the coefficient for a dichotomy gives us the mean difference between groups, but now we get the difference with other predictors controlled.

The same qualification of the bivariate interpretation applies for standardized variables. Suppose we use standardized age and years of education to predict standardized income, and we get

$$Z_{Income} = 0 + .05(Z_{Age}) + .30(Z_{Years\ of\ Education})$$

This means that one SD's increase in age gives, on average, .05 SDs' change in income, with education constant, and that one SD's change in years of education will give, on average, .30 SDs' change in income, holding age constant. (Recall that, in BR, the intercept is always zero for standardized variables. The same is true in MR, for the same reason.)

Table 14.1: Comparing Bivariate and Multiple Regression

	Bivariate	Multiple
Principle for defining equation	Least Squares [Minimize $\sum (y_i - \hat{y}_i)^2$]	Least Squares [Minimize $\sum (y_i - \hat{y}_i)^2$]
Characteristics of regression line	Neither over- nor underestimates, on average [$\sum (y_i - \hat{y}_i) = 0$]	Neither over- nor underestimates, on average [$\sum (y_i - \hat{y}_i) = 0$]
	Passes through centroid $(\bar{x}, \bar{y})$	Passes through multivariate centroid $(\bar{x}1, \bar{x}2 \ldots \bar{x}n, \bar{y})$
Measures of predictive success	SEE ($s_{y.x}$) (SD of residuals)	SEE ($s_{y.x..k}$) (SD of residuals)
	r^2	R^2 (r^2 between y & $\hat{y}$)
Predictor(s)	x	residuals of xj, having predicted it from other IVs
Interpretation of b's (general)	Average number of units of change in y for one unit in x	Average number of units of change in y for one unit in xj, other xs held constant
Interpretation of b's for dichotomous x	Mean difference between categories	Mean difference between categories, other IVs held constant
Interpretation of βs	Average number of SDs of change in y for one SD in x	Average number of SDs of change in y for one SD in xj, other IVs held constant
Interpretation of b's for dummies representing polytomies	Mean difference	Mean difference between reference category and category represented by the dummy, other IVs held constant

Getting the Right Variables into the Equation

Sometimes when we control for other IVs, a predictor that at first seemed to have considerable impact no longer seems to make much difference. It is also true that sometimes an IV that initially seemed to matter very little makes an important contribution to our prediction.

This raises the question of which variables ought to go into our equations. To answer it conclusively, we would need perfect knowledge of what affects what. Granted the difficulty, we must make our best effort.

We use the residuals of an IV, after predicting it from the other IVs, to predict values of the DV. But the residuals we get depend on the other variables in the equation, and so the predictions we make using a given IV can be affected by our selection of others.

Suppose we have left out an IV which has a genuine effect on the DV, and that the omitted IV is correlated with IVs in the equation. The part of their variation that is correlated with the omitted variable will be used in predicting y, so their b's will be *biased*.

Although the presence of bias is not usually obvious, clear examples do arise. The author once tried to predict grades in an introductory class from a few bits of information at his disposal. Among the predictors was a dichotomy called No_Paper, coded

0 = handed in term paper, 1 = did not hand in term paper.

Another was Not_Tried, which represented the number of marks the student did not try to get, by not handing in an assignment (other than the term paper) or not writing an exam.

For simplicity, we will consider the b's only for these two:

For No_Paper, b = −30.238

For Not_Tried, b = −1.651

Those who didn't hand in the term paper, on average, came out just over 30 points below those who did. Since the term paper was worth only 25 marks, No_Paper plainly picked up something more. Again, those who didn't try to pick up marks for exams or assignments, on average, came out worse by 1.65 marks for every point they didn't go after. One surmise is that the omitted variable whose influence these two were picking up was motivation to do well in the course. This seems likely to have been correlated with both No_Paper and Not_Tried, but to have had additional influence on the final grade. Whether motivation was the (only) omitted variable or not, these two variables had to be getting credit for something beyond what they measured directly.

There will be no bias from an omitted variable if it is unrelated to the IVs in the equation. Its inclusion will cause no systematic change in the residuals for other x-variables, so no bias will occur in the b's. But we have to be lucky for (all the) omitted variables to be unrelated to (all the) included variables. Ordinarily, if we suspect that we have omitted relevant variables we must also suspect that at least some of our b's have been biased, at least to a degree.

Suppose, on the other hand, that we include a predictor that is unrelated to the DV. Then its presence will not cause the residuals for other predictors to be any more or any less correlated with the DV. However, if it is correlated with them it will have another harmful effect. It will reduce the residuals of any with which it is correlated, so we will be predicting the DV with less information than we should have had. Thus our b's will be less stable from sample to sample than they otherwise would be. We will need larger samples to get equally precise results, so our estimates of the coefficients will be *inefficient*.

The moral is that we must do the best we can to get the right variables into the equation; otherwise our coefficients are liable to be biased. We must also try to get irrelevant variables out; otherwise our estimates are liable to be inefficient. Since we cannot be absolutely certain which variables should be present, we must always be aware that our results could be, to a greater or lesser extent, misleading.

In a way, we have returned to a version of the third variable problem we encountered in dealing with conditional tables. There is always the possibility that another variable, if taken account of, will change the relationships among the ones we are considering. We work toward the best possible equation, but must be aware of its possible imperfections.

Interpreting a Table of Regression Results

As we have seen, the general interpretation for b's in multiple regression is that they give us the number of units' change in y we get, on average, for one unit of change in xj, with other predictors held constant. There are three special cases in the examples to be considered:

1. With standardized variables, the b's, now called betas, give us the number of SDs' change we get in y, on average, for an SD of change in x_j, other IVs controlled.

2. For a dichotomy, coded with categories one unit apart, b_j gives the mean difference between groups, other IVs held constant.

3. For a set of dummies, the b's give us the mean differences between the categories they represent and the reference category, other IVs held constant.

To make sense of a regression table, we must know which interpretations apply to which variables. Two are needed for Table 14.2.

The variables are as follows:

- Partner satisfaction: This refers to scores on a scale intended to measure quality of marital (or cohabiting) relationships, with higher scores meaning greater satisfaction with the partner.

- Depression: This refers to scores on the Center for Epidemiological Studies Depression scale, which can range from 20 to 80, with higher scores implying greater depression. Since relationship problems can lead to depression, as well as vice versa, depression scores were taken from a previous interview.

- Financial stress: This refers to the number of forms of financial difficulty the respondent reported for the previous three months. These include, for example, not having enough for food and being unable to pay bills. Scores range from 0 to 3.

- Violence: This is scored 1 if violence by either the respondent or the partner (or both) was reported for the previous year. It is scored 0 otherwise.

Since Violence is a dichotomy, the b for this variable tells us the difference between those who reported violence and those who did not. For the other predictors, the more general interpretation applies: b gives us the units of change we expect in partner satisfaction for a unit of change in the predictor.

Table 14.2: Predictors of Partner Satisfaction

Predictor	b	p	β
Depression	−.142	.000	−.232
Financial stress	−1.008	.029	−.153
Violence	−2.578	.025	−.108
Constant	31.935	.000	
R^2	.120		

In the left-hand column you will see the names of the predictors, followed by "Constant," and "R^2." "Constant" is just another name for what we have been calling "a," which for a given equation is a constant. It gives us the value we predict for the dependent variable when all the predictors take on the value of 0. R^2 is the standard PRE measure giving us the proportion of Variance we can account for with this set of predictors.

The second column provides the b's. That for depression appears much lower than the others, but scale scores range from 20 to 80. One unit of increase implies only a change of −.142 in partner satisfaction, but the difference between someone scoring at the bottom of the scale (20) and someone scoring in the middle (50) will be 30 times as great. 30(−.142) = −4.26, a considerable decline. At the other extreme, the highest b is that for Violence, but this variable, as a dichotomy, allows for only a one-point change. Its coefficient tells us that those who report violence are expected to score 2.578 units lower on satisfaction than those who do not. The b for the final predictor, financial stress, implies a reduction of just over one unit in satisfaction for each of the three financial difficulties reported.

The third column provides the p-values. Each gives us the probability of obtaining a b as large as we have, or larger, by chance. All of the ps are below the level of .05, and by convention can be called statistically significant.

P-values for regression coefficients are based on the t and normal sampling distributions. For small samples, t is appropriate, but as sample size increases the sampling distribution for a regression coefficient approximates the normal. Different rules of thumb appear in the literature, but typically the normal is a reasonable approximation for samples of 100 or larger, provided that we are testing at .05. In this instance, we have a sample of close to 400, so the normal should be reasonable. Larger samples are needed for the normal approximation to serve well for more extreme p-values, particularly if the residuals are strongly skewed.

The final column of the table contains the betas, the coefficients obtained after standardizing the variables. There is no value for the constant here, because for standardized variables the value of the intercept is always zero. There is no value given for R^2 either, as it would just be the same as that reported beside the metric coefficients.

In the sociological journals, if metric b's are presented the betas are usually omitted, although you will see them more frequently in other disciplines, such as psychology. It is standard practice to report the means and SDs of variables employed, and with them it is possible to calculate the betas, so that these are not strictly necessary. Another reason they may be less useful than the metric bs is that, unless we look at the standard deviations of the variables, we do not know quite how large the change implied by a beta is.

Table 14.3 introduces another way of reporting p-values. Here ps are represented by asterisks, as shown under the table.

Attitudes to non-marital cohabitation are rated from 1 to 7, with 7 the most favourable score.

There are six dichotomous predictors:

- Sex, coded 0 = M, 1 = F

- Parents divorced or separated, coded 0 = no, 1 = yes

- Previously cohabited, coded 0 = no, 1 = yes

- Lives in large city, coded 0 = no, 1 = yes

- Lives at home, coded 0 = no, 1 = yes

- Currently cohabiting, coded 0 = no, 1 = yes

Age is represented by dummies, with those under 25 as the reference category. Church attendance refers to the number of services attended in the previous month.

Table 14.3: Determinants of Support for Non-marital Cohabitation, for Unmarried Respondents	
Determinant	b
25–34	−.147
35–44	−.403*
45–54	−.688*
55–64	−1.135**
65 plus	−1.627*
Sex	−.223*
Church attendance	−.336**
Lives in large city	−.451**
Lives at home	−1.429**
Currently cohabiting	.343**
R^2	.687

* p < .05; ** p < .01

The dummies for age categories become progressively more negative as age rises. That is, as we move from the reference category of the under 25s, views of cohabitation become progressively less favourable. As expected, church attendance is linked to more negative attitudes. The remaining predictors are dichotomies, so each gives the difference between two groups, controlling for other predictors. You can work out their meaning at your leisure.

Interaction Terms

We have seen how to interpret b's for single variables. If a predictor is dichotomous, and the categories are scored one unit apart, b gives us the mean difference between groups, with other IVs held constant. If a predictor moves over a range of values, b gives us the predicted change in y for each unit of movement along the scale. Sometimes, though, the effect of one predictor depends on the value of another, and we need to represent this through an "interaction term."

To illustrate, we shall extend an example we have seen before. Let us suppose that we have the equation

$$\text{Income} = 4{,}000 + 4{,}000 * (\text{Years of Education}) - 10{,}000 * (\text{Sex}),$$

where sex is coded 0 = M, 1 = F. This equation can be diagrammed as shown in Figure 14.4.

Figure 14.4: Annual Income by Years of Education and Sex (Hypothetical Data)

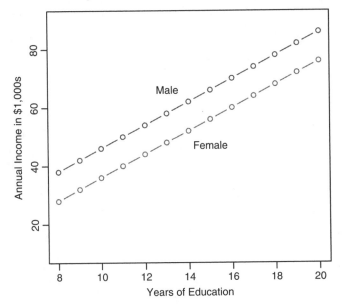

For both males and females, income rises by 4,000 for each additional year of education, but females are always 10,000 below males.

Suppose, though, that the effect of education differs for males and females. Differing slopes can be represented as shown in Figure 14.5.

Figure 14.5: Income by Years of Education Interacting with Sex (Hypothetical Data)

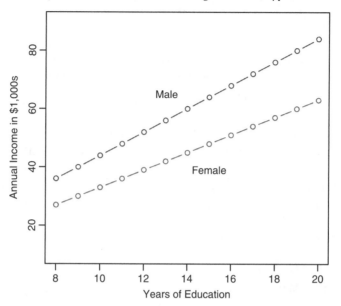

In these hypothetical data, income is rising more rapidly with education for males, so we need a way to build the difference in slopes into the equation. We do this by including a new element, an interaction term. These are produced by multiplying the values for two (or sometimes more) variables together. Here we need to multiply the value for Sex by the value for Years of Education.

The new equation is

Income = 4,000 + 4,000*(Years of Education) – 1,000*(Sex)
– 1000*(Sex)*(Years of Education)

The key difference lies in the final interaction term,

– 1000*(Sex)*(Years of Education)

This term tells us to reduce predicted income by 1,000 for a unit of change in (Sex)*(Years of Education). It only makes a difference for females, though. If the respondent is male, Sex = 0, so the product of sex and years of education vanishes and has no effect on predicted income. If the respondent is female, Sex = 1, so (Sex)*(Years of Education) = (1)(Years of Education) = Years of Education. That is, we now have a second variable representing education completed, which only applies for females. When the respondent is female, for each additional year of education we reduce predicted income by 1,000 below what it would otherwise have been. Since according to this equation it would otherwise have risen by 4,000 for each year of education completed, for females it rises by (4,000 – 1,000), or 3,000 per year.

Concretely, let us see what the equation predicts for a male and a female with 15 years of completed education. For the male,

$$\text{Income} = 4,000 + 4,000(15) - 1,000(0) - 1,000(0)*(15)$$
$$= 4,000 + 60,000 - 0 - 0 = 64,000$$

For the female,

$$\text{Income} = 4,000 + 4,000(15) - 1,000(1) - 1,000(1)*(15)$$
$$= 4,000 + 60,000 - 1,000 - 15,000 = 48,000$$

At this level of education, the difference between males and females is 16,000, but at other levels the gap will not be the same, because income rises with education at a different rate for females.

One consequence of the interaction between education and sex is that we can no longer speak of the effect of education by itself, because its effect depends on sex. Similarly, we cannot speak of the effect of sex by itself, because that depends on education.

For an example of interaction with real data, let us consider some results from Ross and Mirovsky (2006). These authors wanted to understand the link between years of education and depression for males and females. The basic difference, shown in Figure 14.6, was that while depression fell as education rose for both sexes, it fell more rapidly for females.

Figure 14.6: Depression Score by Years of Education

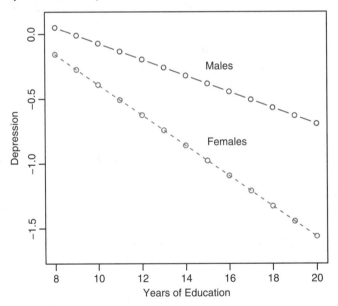

From their table, we can write the equation

Depression = .937 + −.006*(Age) − .140*(White) + .231*(Sex) − .062*(Years of Education) − .055(Sex)*(Years of Education)

Here we need only see how the terms for sex, education, and their interaction work. Sex was coded: 0 = M, 1 = F. Years of Education just represents years completed. For Ms we have

$$.231(0) - .062*(\text{Years of Education})$$

$$- .055(0)*(\text{Years of Education}) =$$

$$- .062*(\text{Years of Education})$$

For Ms, of the terms for sex, education, and their interaction, only the term for education matters, implying a decline of .062 for each year of additional education.

For Fs, we have

$$.231(1) - .062*(\text{Years of Education})$$

$$- .055(1)*(\text{Years of Education}) =$$

$$.231 - .117*(\text{Years of Education})$$

For females, the basic term for sex, the term for education, and the term for their interaction all make a difference to predicted levels of depression. The effect of education for females is given by the term for education and the interaction between sex and education. The two taken together imply that that the decline is more rapid for females, at $-.062 - .055 = -.117$ for each year of additional education.

As always with interactions, we cannot say that education has such and such an effect by itself, because its effect depends on sex. Similarly, the effect of sex depends on education.

An Example with Two Dichotomies

The previous examples involved one dichotomous and one discrete-continuous variable. Combinations of this kind are common in the literature, as are combinations of two dichotomies. Here, we shall consider the joint effects of two of them, each of which can be considered as nominal:

Franco, coded 1 for francophones, 0 for others; and

Sing_par, coded 1 for single-parent households, 0 for others.

They are used in an equation predicting monthly income, written

$$\text{Monthly Income} = 3018 + 1097*(\text{Franco}) - 1447*(\text{Sing\_par})$$

$$- 884*(\text{Franco})*(\text{Sing\_par})$$

The interaction term will disappear for non-francophones, because for them Franco = 0. (It will also disappear for two-parent families, since for them Sing_par = 0). The interaction term will equal 1 only for francophone single parents. The coefficient for the term, which only applies to them, is 884. This tells us that the difference between one- and two-parent households is 884 greater for francophones than for others.

From the equation we can, if we wish, work out predicted incomes for four categories:

1. Francophone two-parent households, for whom we have

$$3018 + 1097(1) - 1447(0) - 884(1)*(0) =$$

$$3018 + 1097 - 0 - 0 = 4115 ;$$

2. Francophone one-parent households, for whom we have

$$3018 + 1097(1) - 1447(1) - 884(1)*(1) =$$

$$3018 + 1097 - 1447 - 884 = 1784 ;$$

3. Other two-parent households, for whom we have

$$3018 + 1097(0) - 1447(0) - 884(0)*(0) =$$

$$3018 + 0 - 0 - 0 = 3018 ; and$$

4. Other one-parent households, for whom we have

$$3018 + 1097(0) - 1447(1) - 844(0)*(1) =$$

$$3018 + 0 - 1447 - 0 = 1571$$

Using these results, we see that for francophones, the difference between two- and one–parent households is

$$4115 - 1784 = 2331$$

For others, the difference is

$$3018 - 1571 = 1447$$

The difference between two- and one-parent households is greater for francophones by

$$2331 - 1447 = 884 , \textbf{which is the value of the interaction term}$$

The bulk of the interactions in the literature are of the two kinds illustrated. However, you will sometimes see an interaction between two discrete-continuous variables. In some exploratory work, the author noticed that older workers tended to have higher job satisfaction than the younger, and that effect of the social standing of the job seemed to matter less for older workers.

The standing of the job was assessed by a Blishen scale (after Bernard Blishen, who created a series of measures over three decades). Rising scores imply progressively "better" jobs. Figure 14.7 is a graph showing how age and Blishen scores affected satisfaction. When interactions involving discrete-continuous variables are shown in 2d, typically only a few lines are drawn, as in Figure 14.7.

Figure 14.7: Job Satisfaction by Blishen Score and Age

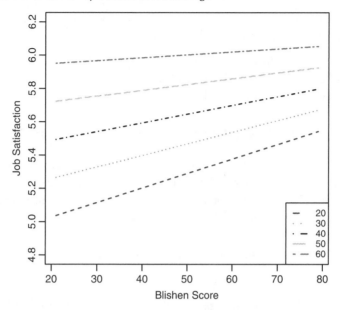

It would be possible to draw a line for each year of age, but the graph would be very cluttered, and the point that Blishen scores were of decreasing importance to satisfaction as age rose is made by the lines shown.

Setting Up a Regression Table

How to set up a regression table will be illustrated with a table based on Nakhaie and Arnold (2010), presented as Table 14.4. As illustrated above, these tables appear in varying formats. The precise format required varies with discipline, and even from journal to journal. Here, we will use the most compact, as other formats can be employed readily once how to set up the most basic is mastered. As for mean and SD tables, and for crosstabulations, we need to choose a monospace font to be able to align the decimal points. Some common ones are Andale Mono, Bitstream Vera Sans Mono, Courier, Courier New, DejaVu Sans Mono, FreeMono, Lucida Console, and Monaco. We can choose based on availability, legibility, and appearance.

The dependent variable is changes in functional health status, coded so that 1 equals perfect health and 0 is equivalent to death. Changes over a four-year period have been calculated, and regressed on four predictors, coded as follows:

- Female = 1 if R is female, 0 otherwise

- Age is given in years above 64. (Age has been truncated because the trend line is flat below 64.)

- Perceived love = 1 if Rs say there are others they love and who love them, 0 otherwise

- Food insecurity is coded 0 to 3, where 0 means there was always enough money for food and 3 means R often went without enough food.

The table title typically does not name the independent variables, but refers to them generically, often as "predictors" or "determinants." The dependent variable is named. The title is centred over the table, and bolded or italicized. Here, in Table 14.4, we have

Table 14.4: Determinants of Change in Health Status

Next we need column headings, one for the IVs and one for the coefficients. The heading for the IVs will often be "Predictors," "Determinants" or "Independent Variables." This heading will be left justified, while the heading for the coefficients will be centred over its column. These headings will be bolded or italicized, whichever has been done with the table title.

Table 14.4: Determinants of Change in Health Status

Predictors	**b**

Now, on the left, we add the names of the predictors. These are neither bolded nor italicized. They must be left justified.

Table 14.4: Determinants of Change in Health Status

Predictors	**b**
Female	
Age	
Perceived love	
Food insecurity	
Intercept	

Next, we add the b's, centred under the column heading, with the decimal points aligned.

Table 14.4: Determinants of Change in Health Status

Predictors	**b**
Female	−.009
Age	−.019
Perceived love	.036
Food insecurity	−.019
Intercept	.036

P-values are sometimes placed in a separate column, and sometimes indicated by asterisks or other symbols. Using a separate column allows more detail to be presented, but also requires more space. Here we will follow the more common practice of using superscripted asterisks. When these are used, it is common to distinguish .05, .01, and .001 levels, with one asterisk for .05, two for .01, and three for .001.

Table 14.4: Determinants of Change in Health Status

Predictors	b
Female	−.009***
Age	−.019***
Perceived love	.036***
Food insecurity	−.019***
Intercept	.036*

To make sure there is no misunderstanding, the meaning of the symbols is indicated below the table, often below a line. Specific publishing outlets often have their own guidelines about the length or style of the line.

We have also added R^2 to the table, below the names of the predictors, but above the explanatory note. If the SEE is presented, it goes on a separate line above or below R^2. Here an extra line has been placed between the coefficients and R^2. A visual break here is common enough, but not obligatory.

With these elements added, the table is complete.

Table 14.4: Determinants of Change in Health Status

Predictors	b
Female	−.009***
Age	−.019***
Perceived love	.036***
Food insecurity	−.019***
Intercept	.036*
R^2	.084

*p < .05; **p < .01; ***p < .001

Creating an Equation

We do not often present regression results in the form of an equation, but it is easy to create one from the same data we put into a table. We just have to plug the variable names and their coefficients into an equation of the form

$$\hat{y} = a + b_1x_1 + b_2x_2 \ldots + b_{nx}n$$

We can begin by substituting the variable names, or short forms for them. Then we can add the coefficients. Since there is no easy way to put a hat over the name of the dependent variable, we will skip that. We get

$$\text{Change} = a + b_1*(\text{Female}) + b_2*(\text{Age}) + b_3*(\text{Love}) + b_4*(\text{Food})$$

$$= .036 - .009*(\text{Female}) - .019*(\text{Age})$$

$$+ .036*(\text{Love}) - .019*(\text{Food})$$

Graphs Presenting Regression Results

Tables are more common in journals because they are compact and provide precise figures, but graphs may be more readily understood by an audience unfamiliar with regression. Figure 14.8 shows two graphs taken from a prediction of household income.

The predicted value for a case depends on the scores on several IVs. We can show how well the prediction went overall, and can also show what each predictor suggests. To do the latter, we can set the values of IVs other than those being graphed to their means. Then we get trend lines for each that run through the centroid.

Figure 14.8: Household Income by Predicted Income and Prior Income

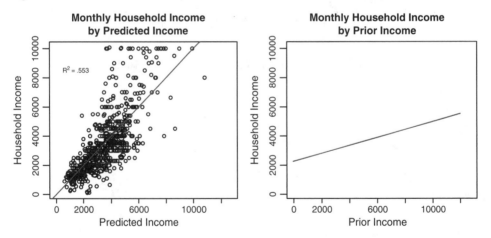

The graphs in Figure 14.8 have been set up with the same range of values on their y-axes, so the impact of all the IVs, then the effect of an individual predictor, may be compared.

In the left-hand graph, we see not just a trend line, but departures of individual cases from their predicted values. R^2 is shown, with a value of .553, indicating that the overall reduction is better than most in the social sciences.

The trend line for the effect of income at prior interviews is shown on the right. Of course this effect is not as great as that of the full set of predictors. (These include the respondent's education, the number of adults with full-time jobs, and the number with professional, managerial, or technical jobs.) Moving from one extreme of prior income to the other, the predicted value for monthly income goes from just over 2,000 to over 6,000 per month.

The Special Case of Analysis of Variance (ANOVA)

We have seen that dichotomous variables can be used as predictors, and that categorical polytomies can be represented by sets of dummy variables. We have also seen that we can use more than one categorical variable in a single equation. Analysis of Variance (ANOVA) is a special case in which the predictors are all categorical.

One-Way ANOVA

Forms of ANOVA are distinguished by the number of predictors used. In the sociology journals today, one-way ANOVA, which is equivalent to regression using a set of dummy variables, is the most common form, but two-way ANOVA is very common in psychology.

We might, for example, want to see how far ethnic identity goes in predicting monthly income. One data set contains four categories:

- Canadian-Born Anglophones;

- Canadian-Born Francophones;

- Natives; and

- Others.

Their relative standing can be displayed graphically, as shown in Figure 14.9, where the line across the centre of the graph represents the overall mean for the sample, the "grand mean" as it is called in ANOVA.

Figure 14.9: Mean Monthly Income by Ethnicity

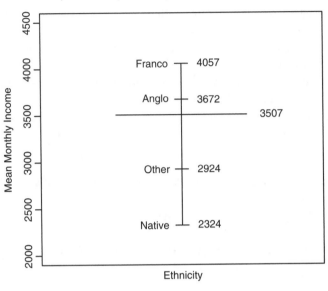

Using the "Other" category as the reference, the corresponding regression equation turns out to be

$$\text{Income} = 2924 + 748*(\text{Anglo}) + 1133*(\text{Franco}) - 600*(\text{Native})$$

R^2 is .061 and the SEE is 2078.

If we run ANOVA, the output will typically provide the means for the groups. If we run regression, we can work out the means from the corresponding equation. Each group with its own dummy will take the value of the intercept plus the value of its dummy. The Others, with no term of their own, take the value of the intercept. In this case, the means are as follows:

for the Anglophones: $2924 + 748 = 3672$;

for the Francophones: $2924 + 1133 = 4057$;

for the Natives: $2924 - 600 = 2324$; and

for the Others: $2924 + 0 = 2924$.

Plainly, though, if our interest lies in the means it is simpler to use a one-way ANOVA routine. These often provide a table of the differences among the means as well.

Two-Way ANOVA

Another widely used form, two-way ANOVA, is also a special case in that the predictors are uncorrelated. This form is primarily used in experiments, with each category of interest equally represented. For example, in a hypothetical examination of how hard Grade 9 students were working, subjects could be assigned as follows.

	Control	Treatment	All Cases
M	25	25	50
F	25	25	50
Total	50	50	100

Being in the treatment group and sex are uncorrelated—neither can be predicted from the other. If there were more predictors, the numbers would be set up to balance all the groups created.

Apart from using categorical predictors, set up to be uncorrelated, two-or-more way ANOVA is standard regression. However, the output from ANOVA programs differs from that of regression routines. In the hypothetical student effort study, the variables were:

Workhard, representing teacher ratings of student effort, coded 1 to 7, with higher scores implying more effort;

Sex, coded 0 = M, 1 = F; and

Treated, coded 0 = no, 1 = yes

Mean scores for boys and girls, and for treatment and control groups, can be presented graphically, as in Figure 14.10. The long line across the centre of the graph represents the grand mean.

Figure 14.10: Mean Teacher Ratings by Type of Site and Sex of Child

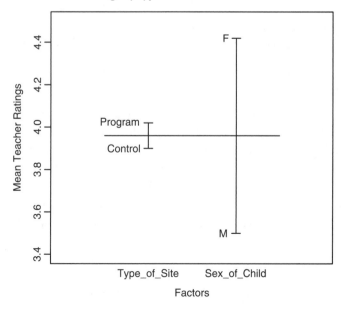

Solving the regression equation

$$\text{Workhard} = a + b_1*(\text{Sex}) + b_2*(\text{Treated})$$

$$= 3.44 + .92*(\text{Sex}) + .12*(\text{Treated})$$

R^2 was .058 and the SEE was 2.15.

Girls were, on average, rated as working harder, by .92, but the treatment effect, at .12, was, for practical purposes, trivial.

Although at one time two-way ANOVA routines did not typically report R^2 and the SEE, this is now common. They still do not typically provide equations, but they do provide predicted means for each set of cases, as illustrated in the table below.

	Control	Treated	All Cases
M	3.44	3.56	3.50
F	4.36	4.48	4.42
Total	3.90	4.02	3.96

The difference between Fs and Ms for all cases combined is

$$4.42 - 3.50 = .92 \,,$$

the same as the b for Sex in the equation. Similarly the difference between treated and control cases, at

$$4.02 - 3.90 = .12 \,,$$

is the same as the b for Treated.

If we want the predicted means, we can again calculate them from a regression equation, but it is simpler to rely on an ANOVA routine.

Interactions

In looking at conditional tables, none of which were based on experiments, we have seen that the effect of one variable often depends on that of another. We have seen that the same may be true in regression analyses. In experiments, we often want to see whether this is true. If it is, and we have used ANOVA, we speak of "interaction" between predictors. This possibility is represented in "interaction plots," as illustrated below.

Teacher ratings for boys are about the same whether they are in the control or the treatment group, but girls are doing noticeably better in the treatment group. We could have built the interaction between sex and group into our equation, but the difference between boys and girls is small enough to be readily attributed to chance, so in this example we really only need to be sure we are clear on the elements in the graph.

In plots of this kind, two groups or settings of interest—often the experimental and control groups—are distinguished on the x-axis. The outcome is represented on the y-axis, and trend lines are labelled to distinguish two further groups or situations. In Figure 14.11, control and demonstration groups are distinguished on the x-axis, teacher ratings are represented on the y-axis, and Ms and Fs are represented by separate trend lines.

Figure 14.11: Teacher Ratings of Student Effort by Type of Site and Sex of Child

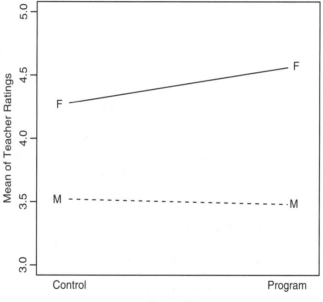

A second graph, in Figure 14.12, is based on Correll's (2004) experiment with male and female undergraduates. Half were exposed to a statement suggesting that males tended to have greater ability in the type of task they were to perform, and half were not. Correll predicted, correctly, that males would rate their performance higher when they had been exposed to the message than if they had not.

Figure 14.12: Effect of Statement about Greater Male Ability on Self-Rating of Test Performance

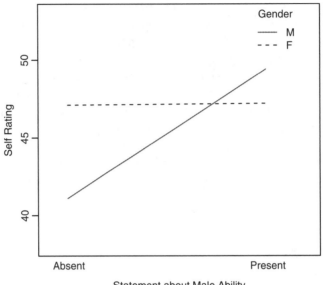

In this graph, the x-axis identifies not two groups, but two experimental conditions, in only one of which the statement on male ability was presented. The two groups, males and females, are identified by different line styles, one solid and one dashed. Clearly, the mean scores for males rose when they were exposed to the statement, but it had no appreciable effect on females.

To represent the interaction between sex and treatment in an equation we need an "interaction term." As we have seen above, these are created by multiplying the scores of the two independent variables involved. Here we can multiply

$$(\textbf{Gender}) \times (\textbf{Statement})$$

The value of the product of course reflects the coding of the two variables. If they are coded

male = 1, female = 0, and

statement absent = 0, statement present = 1,

the value of the interaction term becomes as follows:

for females, without the statement $0 \times 0 = 0$;

for females, with the statement $0 \times 0 = 0$;

for males, without the statement $\qquad$ $1 \times 0 = 0$; and

for males, with the statement $\qquad$ $1 \times 1 = 1$

The interaction term equals 1 for males exposed to the statement, and 0 for others. The model can be expressed in an equation:

$$\text{Self-rating} = a + b_1*(\text{Gender}) + b_2*(\text{Statement} + b_3*(\text{Interaction})$$

The final term, involving the interaction, will disappear for everyone but males exposed to the statement, because for everyone else the value of the interaction is zero.

The solution to the equation, which can be worked out from information provided by Carroll, is

$$\text{Self-rating} = 47.1 - 6*(\text{Gender}) + .1*(\text{Statement}) + 8.2*(\text{Interaction})$$

Predicted values based on the equation are as follows:

for females, without the statement,

$$47.1 - 6(0) + .1(0) + 8.2(0) = 47.1 \ ;$$

for females with the statement,

$$47.1 - 6(0) + .1(1) + 8.2(0) = 47.2 \ ;$$

for males, without the statement,

$$47.1 - 6(1) + .1(0) + 8.2(0) = 41.1 \ ; \text{ and}$$

for males with the statement,

$$47.1 - 6(1) + .1(1) + 8.2(1) = 49.4$$

These are the values plotted in Figure 14.12.

As a primary interest in many experiments is whether or not an interaction is present, plots of the kind we have just seen are common in the experimental literature. Each, if we wish, can be expressed as a regression equation of the form we have seen. Let us review another example.

In a study by Janssens et al. (2010), male students were asked to recall products they had viewed, some of which were both conspicuous and expensive—e.g., large houses and high-performance sports cars—and which were called "status products." The men were divided randomly into groups that viewed and recalled the products in the presence of a female experimenter dressed either plainly or more glamorously. The hypothesis was that unattached men, exposed to an attractive female in a "notice-me" outfit, would become more interested in products that would show their economic status. The results are presented in Figure 14.13.

Figure 14.13: Effects of F's Style of Dress and M's Availability on N of Status Products Recalled

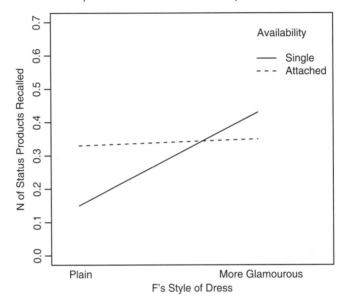

As anticipated, the female's style of dress made no appreciable difference to men who were already attached, but mattered for those who were not.

These results can be put into an equation once we have codes for the predictors. Let us code

Dress: 0 = plain, 1 = more glamorous; and

Availability: 0 = attached, 1 = single

Using these codes, single men paired with more glamorously dressed women will score $1 \times 1 = 1$ on the interaction term. All others will score 0.

On this basis, the equation will turn out as

$$\textbf{Products} = \textbf{a} + \textbf{b}_1\textbf{*(Availability)} + \textbf{b}_2\textbf{*(Dress)} + \textbf{b}_3\textbf{*(Interaction)}$$

$$= \textbf{.35} - \textbf{.18*(Availability)} - \textbf{.02*(Dress)} + \textbf{.28*(Interaction)}$$

For single Ms and more glamorous Fs, we get

$$\textbf{.35} - \textbf{.18(1)} - \textbf{.02(1)} + \textbf{.28(1)} = \textbf{.43}$$

You can work out the other predictions at your leisure, or rely on the interaction plot for the basic message of the study.

Analysis of Covariance

In experiments it is often helpful to control for a variable that is not categorical. In the analysis of student effort above, it might have been useful to control for grades in the year before the treatment took place. Ordered variables used as controls are typically called covariates, and the statistical procedure is renamed Analysis of Covariance (ANCOVA). It need not be considered further here, since it is multiple regression under another name.

The F Distribution in Regression and ANOVA

The F distribution was alluded to earlier, in Chapter 6, on sampling distributions. There, it was pointed out that the discussion would be left to this section. In the social sciences, the F distribution appears primarily in regression and ANOVA. Significance tests for comparison of means in ANOVA are based on it. It is also used in regression when we want to check the significance of a set of predictors rather than testing them one at a time.

Like t and chi-square, the F distribution is really a family of distributions. Like t, the F distribution is a ratio of two others. It is a ratio of two chi-squares, each divided by its degrees of freedom. Formally,

$$F = \frac{\chi^2 / (DF)}{\chi^2 / (DF)}$$

F distributions are continuous and unimodal, with no values below zero. In practice, they usually appear right-skewed, but as both DF increase they appear more and more symmetric. Beyond these generalities, their shape varies greatly, because it depends on two underlying distributions, each with varying DF. To distinguish F distributions, we have to provide both. We put those for the numerator first, so, for example, if there are three DF in the numerator and 96 in the denominator the distribution is F(3,96). Figure 14.14 shows four Fs with differing DF.

Figure 14.14: Four F Distributions

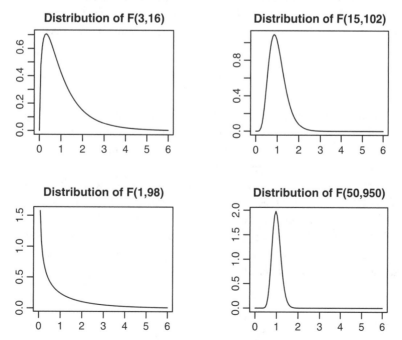

Recall that in defining a regression line, we wish to minimize the sum of squared errors,

$$\Sigma (\hat{y}_i - y_i)^2$$

This quantity may be referred to as the Error Sum of Squares (ESS). If the differences between $\hat{y}_i$ and y_i are normally distributed, we are summing squared normal deviates, so the ESS is distributed as chi-square.

When we don't know how y is linked to its predictors, we minimize our squared errors in predicting y by guessing that each case will take on its mean value. In algebraic notation, we take

$$\Sigma (y_i - \overline{y})^2$$

This quantity may be referred to as the Total Sum of Squares (TSS). If the differences between y_i and $\overline{y}$) are normally distributed, we are again summing squared normal deviates, so the TSS is distributed as chi-square.

The difference between the total sum and the error sum of squares is our reduction in squared errors, which may be referred to as the regression sum of squares (RSS). Assuming the TSS and ESS are chi-square variables, the RSS will be as well.[3]

To create the F distribution used in regression, and in its special case of ANOVA, we use the sum of squares due to the model, the RSS, in the numerator, and the error sum of squares, the ESS, in the denominator. We have

$$\frac{\text{RSS}}{\text{ESS}}$$

On the assumptions about normality stated above, each of these is distributed as chi-square.

We divide each by its degrees of freedom. For the numerator, DF equals the number of predictors used, which we shall denote as k. For the denominator we take N − 1 − k. So

$$F = \frac{\text{RSS} / k}{\text{ESS} / (N - 1 - k)}$$

If our prediction is about as good as might be expected by randomly choosing predictors, the ratio will be close to one. If our prediction is a good bit better than might be expected on the basis of chance F will be large. Software will tell us how likely we are to have obtained a given value of F, or a more extreme one, by chance.

In ANOVA we often test several group means at once, with the null hypothesis that the groups do not differ. Any given mean, if higher or lower than the others, counts as evidence against the null. However, the test is almost always one-tailed, in the sense that only results in the right tail of the F distribution are taken as significant. These imply that our explained variation is large in relation to what we would likely have achieved by chance. Here, we will return to the mean incomes of cultural groups.

The group means are, in descending order:

Francophone: 4057 (SD = 2078)

Anglophone: 3672 (SD = 2518)

Other: 2924 (SD = 1282)

Native: 2324 (SD = 1569)

Knowing that the total sample came to 618, we might well doubt the null, but we need a formal test. Recall that when we set up dummy variables we need g − 1, so here we need 4 − 1 = 3, each to be used as a predictor. Thus k, the DF for the numerator, is 3. For the denominator df = N − k − 1, so here we have 618 − 3 − 1 = 614. We obtain

$$F(3,614) = 13.338, p < .0001$$

We plainly cannot be satisfied with the null. The graph in Figure 14.15, like the p-value, shows that F is in the *extreme* right tail of the sampling distribution.

Figure 14.15: A Test of the Null That Ethnicity and Income Are Independent, F(3,614)

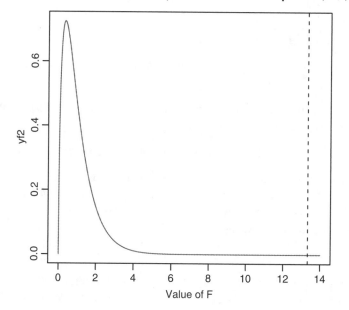

Summary

Typically we need more than one predictor to account for a variable of interest, either because a single predictor does not do a very good job, or more often, because our theories suggest that we must try several. Multiple regression allows us to estimate the effects of several simultaneously, and gives us the effect of each independent of the others. We obtain two measures of how well we are doing, R^2 and the standard error of estimate.

Sometimes the slopes are different for one group than for another. We deal with this through interaction terms.

A special case of regression is Analysis of Variance (ANOVA), used to compare the means of specific subgroups. In sociology we most often see one-way ANOVA (with a single predictor), but in more experimentally oriented disciplines, such as psychology, we often see two-way ANOVA. In this form, we are often very interested in interactions, in situations in which the effect of one predictor depends on the value of another.

The sampling distribution of ANOVA, and for regression when we want to test the effects of more than one variable at a time, is the F distribution, whose properties we have seen.

Review Questions on Multiple Regression and ANOVA

1. What principle do we use as the basis for selecting an equation in *multiple* regression? What are three advantages of choosing an equation in this way?

2. What is the difference between r^2 and R^2?

3. In multiple regression, how do we obtain coefficients that show the effect on an IV, independent of other IVs?

4. Why don't we just run a series of bivariate regressions when we have a set of predictors we want to check out, rather than moving to MR?

5. In the equation $y = a + b_1(X_1) + b_2(X_2)$, what does a tell us? What does b_1 tell us?

6. Suppose we want to predict grades in this course on a scale of 0 to 100, and we want to use three IVs: grades in high school math (averaged over courses), hours per week of work on the course, and year at university, coded 0 = third year, and 1 = fourth year. Suppose we obtain the following equation:

   Grade = 4 + .7*(High School Grades) + 1*(Hours/week) + 5*(Year)

   a) From this equation, what grade would we predict for someone whose high school grades averaged 80, who works 10 hours per week and is in the third year at university?

   b) Suppose that for this equation R^2 is .50, and the standard error of estimate is 5.0. How do we interpret these two measures?

c) Suppose the errors turn out to be normally distributed. How can we use the standard error of estimate to determine the range in which the smallest 95% of them lie?

7. The instructor in another course wanted to predict grades on the final examination, using midterm grades, term paper grades, and identity of the grader as IVs. The variables were as follows:

Final Exam: scored 0–100;

Term Paper: scored 0–100;

Midterm: scored 0–100; and

Grader: coded: 0 = grader A, 1 = grader B

He obtained the following results:

Final Exam = 10.701 + .286(Term Paper) + .763(Midterm) − 6.197(Grader)	
Predictor	**b**
Term paper	.286
Midterm	.763
Grader	−6.197
Constant	10.701
SEE = 8.418	
R^2 = .525	

a) What does the b for Midterm tell us? The b for Grader?

b) What does the R^2 we have tell us? The SEE?

c) The instructor had wondered whether he should look at some papers to see if the graders had handled things the same way. Do you think these results suggest he should have? Why?

8. In multiple regression, what weakness will our estimated coefficients have if (a) we omit an IV which should be in the equation, or (b) we include an irrelevant IV in the equation?

9. What type of term do we introduce to a regression equation when the effect of one predictor depends on the value of another?

10. In what two respects is two-way ANOVA a special case of regression? In what way is one-way ANOVA a special case?

11. What are we likely to find in one-way ANOVA results that we do not find in standard regression output?

12. In ANOVA, when do we get an interaction? How do we represent an interaction in an equation?

13. What is an interaction plot? Make up some data and draw the corresponding plot.

14. What is ANCOVA?

Notes

1. We have seen above that the only exception liable to be seen in the social sciences is the t distribution with fewer than five degrees of freedom.

2. In fact R^2 equals r^2 between y and $\hat{y}$ rather than between y and any of its individual predictors.

3. When we subtract one variable distributed as χ^2 from another distributed as χ^2, the resulting third variable is also distributed as χ^2. Here, in taking TSS − ESS = RSS, we are subtracting one χ^2 variable from another, hence getting a third.

Path Analysis

Learning Objectives

In this chapter, you will learn

- how we can represent a network of causes in a diagram;

- how we can estimate the impact of each cause we have placed in the diagram using regression; and

- how we examine the way in which one variable affects another indirectly, by working through other variables.

A Famous Path Model

Path analysis was developed by a geneticist, Sewell Wright, before the end of the First World War, but was little used in the social sciences until the 1950s and 1960s. Otis Dudley Duncan's paper "Path Analysis: Sociological Examples," appearing in the *American Journal of Sociology* in 1966, drew wide attention to its potential.

We shall examine what may have been the most famous path model in sociology, presented by Blau and Duncan (1967). Literally hundreds of papers have presented variations and extensions of their basic model. Although their topic has since been examined with much more complex data, the nature of path models is well illustrated in this one, and everything works out tidily to three decimal digits. A diagram of their model appears in Figure 15.1.

Figure 15.1: Blau and Duncan's Model of Occupational Attainment

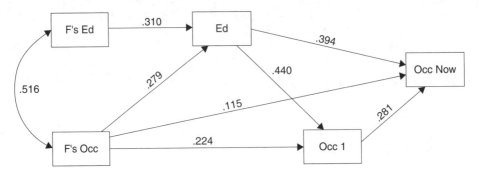

In this model, in Figure 15.1, on the left we have "F's Ed," father's education, and "F's Occ," father's occupation. To the right of father's education, we have education of the respondent, measured in years completed. Below it and to its right, we have "Occ 1" the respondent's first occupation after completing full time education. On the far right, we have "Occ Now," the respondent's occupation at the time of the survey.

Blau and Duncan's intention was to assess the extent to which the American occupational structure was "open," that is, the extent to which individual talent and enterprise enabled advancement, independent of things over which someone has no control, including family background. As part of such an assessment, we must look at the impact of parental education and occupational standing.

Occupational standing was estimated by using a weighted average of education (a primary prerequisite for many "good" jobs) and income (an outcome of many "good" jobs).

The arrows indicate the flow of causality. So, for example, we find that F's Ed affects respondent's education, which in turn affects both Occ 1 and Occ Now. Respondent's education also has an effect on current occupation over and above its effect through Occ 1. Other effects will be examined below.

The attractions of path analysis are several:

1. path diagrams display causal networks in an easily understood form;

2. they allow us to make explicit quite complex models, which once worked out can be tested; and

3. given the r between two variables, we can work out

    a) how much of their association is due to common causes;

    b) how much is due to a direct causal effect of one on the other;

    c) how much is due to indirect effects of one on the other, working through intermediate variables; and, where it is appropriate,

    d) how much is due to effects we are not going to look at in causal terms.

In the diagram, the ultimate dependent variable, current occupation, is found on the far right, parental characteristics on the left, and intervening variables (education and first job) between them. In its setup the diagram illustrates common practices in setting up path diagrams.

Setting Up Path Diagrams

You will see variations, but the following practices are widely adhered to:

1. Cause-and-effect relationships are represented by single-headed arrows, pointing from cause to effect. Here, for example, father's education and occupation send arrows to respondent's education, because the former are seen as causally prior.

2. Each unanalyzed association (for which cause and effect are not considered) is represented by a curved, two-headed arrow. Here the link between father's education and occupation is of this form because Blau and Duncan took these as givens, as background variables whose influence was to be traced, but whose causal linkage need not be analyzed for their purposes.

3. Reciprocal causal links are represented by two single-headed arrows, pointing back and forth between the two variables involved. Here no reciprocal effects are included.

4. Usually cause-and-effect is shown by arrows pointing rightward. (You will see numerous exceptions.)

5. Usually the names of measured variables are placed in rectangular boxes. Hypothetical variables, measured without error, are typically placed in ovals or circles.

6. Coefficients representing the impact of one variable on another are typically placed above the lines linking them. This practice has been followed here. If the arrows are not horizontal, coefficients often go beside them. In this diagram, Pearson's r was placed beside the double-headed arrow, which curves upward.

Equations

Each of the coefficients, except for the correlation between father's education and occupation, is taken from a regression equation. A glance at the diagram will make it clear which equations have been run. Each independent variable must be the origin of an arrow, and each dependent variable must be the destination of one or more. For example, current occupation is the destination of arrows from initial occupation, education and father's occupation, so there must have been an equation predicting current occupation from those three variables. Similarly, Blau and Duncan must have run an equation predicting initial occupation from father's occupation and respondent's education. Their final equation must have predicted respondent's education from father's education and occupation.

In this example, the coefficients in the diagram come from equations employing standardized variables. Often metric coefficients are presented instead, but here the betas will be helpful in illustrating the decomposition of a correlation.

Recall that one interpretation of r is that it tells us how many SDs of change we expect in one variable for an SD of change in the other. That being so, if we are to decompose it, we must work with standard deviations as our units, and so we use the betas, which tell us how many SDs of change in an outcome we expect for an SD of change in a predictor (other variables controlled).

Decomposing a Correlation

If we have used standardized variables, we can decompose the r between one variable and another, causally downstream, into

1. a direct effect, operating with no intervening variables;

2. indirect effects, working through intermediate variables;

3. unanalyzed effects, which involve variables whose causal link is unknown or not to be analyzed; and

4. spurious effects, in which the two key variables share a common cause or causes.

If the model is correct, the sum of these will equal the original r.

We can work out the decomposition ourselves for simple models. We shall illustrate with the r of .405 between father's occupation and respondent's current occupation. This was obtained prior to the path analysis, and is taken as a given, which must be understood.

1. A direct effect is simply the beta for a path running directly from the causally prior variable to its effect. This will just be the beta from the equation in which the first variable is a predictor of the second. Here the beta is the .115 on the line running through the centre of the diagram.

2. To obtain indirect effects, we multiply the betas along each path linking the two variables (without going backward). We have three paths:

   • one from F's Occ through education (Ed) to occupation at the time of the survey (Occ Now);
   • a second from F's Occ through first occupation (Occ 1) to Occ Now; and
   • a third from F's Occ through Ed to Occ 1 to Occ Now.

We multiply as follows:

F's Occ > Ed > Occ Now	$(.279)(.394) = .110$
F's Occ > Occ 1 > Occ Now	$(.224)(.281) = .062$
F's Occ > Ed > Occ 1 > Occ Now	$(.279)(.440)(.281) = .035$
	$\overline{.207}$

The sum of the indirect effects, at .207, is approaching twice the direct effect of .115. Further, the first of the indirect effects is greater than the other two put together.

3. To obtain unanalyzed effects, we multiply the rs and the betas along paths linking the two variables. Here we have only one source of unanalyzed effects, the link between father's occupation and father's education. This link results in two pathways, for which the multiplication goes as follows:

F's Occ > F's Ed > Ed > Occ Now	$(.516)(.310)(.394) = .063$
F's Occ > F's Ed > Ed > Occ 1 > Occ Now	$(.516)(.310)(.440)(.281) = .020$
	$.083$

4. To obtain spurious effects (effects of common causes), we go backward to a common cause then forward to its second effect rather than along a curved arrow to a correlated variable, then forward. Here there are no spurious effects.

5. Finally, we sum the direct, indirect, unanalyzed, and spurious effects. If the causal model is correct, they should equal Pearson's r. Here we obtain

direct effect	.115
indirect effects	.207
unanalyzed effects	.083
	.405

The sum matches the original r between father's occupation and respondent's current occupation.

Of course, path analysis is problematic if we cannot specify causal order, and this may be very difficult. When we can, this approach can be immensely useful. Since the 1960s, it has been extended in multiple ways, in what is called Structural Equation Modelling. The many extensions must be left for another occasion.

Steps in Decomposing a Correlation

1. Create a path diagram. Be sure that each DV has one or more arrows pointing to it, and that each IV sends an arrow to every variable it is expected to influence.

2. Run an equation to predict each DV. To decompose an r, they must be run with standardized variables.[1]

3. The direct effect is simply the beta for a path running directly from one of the key variables to the other.

4. To obtain indirect effects, multiply the betas along each causal pathway linking the two variables whose r is to be decomposed.

5. To obtain unanalyzed effects, multiply the r and the betas along pathways linking the two variables of interest.

6. To obtain spurious effects, go back from your first key variable to the common cause, then forward to the second, multiplying betas (and if necessary, rs) along the way.

7. Sum the direct, indirect, unanalyzed, and spurious effects. If the causal model is correct, they should equal Pearson's r.

A Further Example

Table 15.1 provides the regression results underlying Figure 15.2. In its current form, the diagram includes only + and − symbols to indicate the directions of association. From the table below we can find the figures that could be placed above the arrows, or work out the strength of direct and indirect effects.

	Table 15.1: Determinants of Parental Depression			
	Dependent Variables			
Independent Variables	**Financial Stress**	**Life Events**	**Social Support**	**CESD Scores**
Age		−.057		
Sex				2.214
Single parent		.301		
Married		−.608		
Education	−.028			−.585
Employed full time	−.094		.604	
ln(income)	−.178			−1.463
Contacts with friends			.069	
Neighbourhood activity			.098	
Family functioning			.217	−.432
Financial stress				1.324
Life events				.993
Social support				−.742
Constant	1.166	4.492	24.061	4.013
R²	110	.051	.228	.232

The coefficients in the table are unstandardized, so we cannot use them to decompose a correlation. We can, though, use them to work out indirect effects. Here we shall focus on the equation predicting life events. These are events likely to create stress, e.g. loss of a job, illness in the family, or divorce.

Figure 15.2: Determinants of CESD Scores

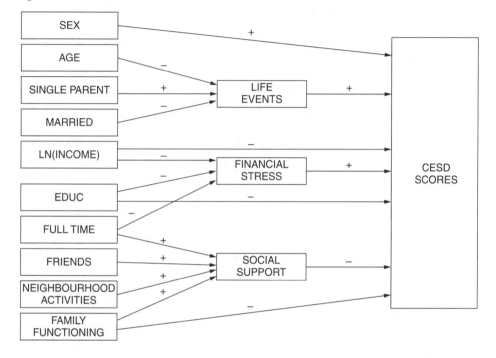

The predictors, as shown above, are age, single parenthood, and being married. For age, the coefficient is –.057, which suggests a slow decline in life events as age, which is coded in 10-year categories, rises. This figure could be placed above the line between the two variables on the graph. The coefficients for single parenthood and being married are contrasted with a reference category of unmarried people who do not have children. For single parents the coefficient is .301, which suggests that single parents, on average have about 3 / 10s more events than the reference category. The coefficient for the married suggests that they have .608 fewer.

To calculate indirect effects on CESD scores, we just have to multiply the b's along the paths. The effect of life events on the CESD is represented by a b of .993. The indirect effects are as follows:

age → life events → CESD	$(-.057)(.993) = -.057$
single parent → life events → CESD	$(.301)(.993) = .299$
being married → life events → CESD	$(-.608)(.993) = -.604$

These effects are expressed in the units of the final dependent variable—that is, in points on the CES Depression scale. Being married reduces scores, on average, by .604 points. By affecting life events, single parenthood raises CESD scores by .299. A 10-year change in age lowers them, on average, by .057. This seems a small effect, but let us recall that this is the change we get for a 10-year shift in age. Going from, say, 25 to 75 has five times this effect, $5(-.057) = -.285$, about the same effect as single parenthood.

Other indirect effects can be worked out in the same way.

Summary

Path analysis, in the form we have considered, is an extension of multiple regression, which allows us to deal with causal chains. We diagram the causal chains we hypothesize, and from the diagram we can read off the necessary regression equations. Given equations in standardized form, we can decompose the r between one variable and another which is causally downstream into

- a direct effect;

- indirect effects, working through intervening variables;

- unanalyzed effects, whose causal nature is not examined; and

- spurious effects.

If we are not interested in decomposing an r, we can still estimate the strength of indirect effects by multiplying the coefficients along the paths.

In the social sciences, we often believe that our topic of interest requires a complex causal analysis. Because this is so, many extensions of the basic path model have been developed, but these must be left for another time.

Review Questions on Path Analysis

1. Suppose that we have four variables, called A, B, C, and D. Suppose that we are not interested in what causes A or B, but we know they are correlated, and we believe that

    C is affected by A and B; and

    D is affected by B and C.

    Draw the corresponding diagram.

2. For the diagram below, what equations should be run? For each, indicate the dependent variable and its predictor(s).

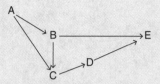

3. Suppose you wanted to decompose the r between B and E in the diagram above. What effects would you have to take account of?

4. If we want to decompose the r between two variables, do we run our regressions with standardized or unstandardized variables? Why?

5. What is a "direct effect"? An "indirect effect"? An "unanalyzed effect"? A "spurious effect"? How do we calculate them?

6. Suppose you see the following diagram, and suppose that, for variables A and E, you know that r = .40. Decompose it into direct and indirect effects.

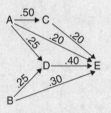

Note

1. Analyses may be run with unstandardized variables, and direct and indirect effects may be calculated either way. It is only in decomposing a correlation that we are restricted to standardized variables. (Recall that for comparison across data sets we have reason to prefer equations based on unstandardized variables.)

16

Logistic Regression

Learning Objectives

In this chapter, you will learn how we adapt regression to deal with a dichotomous dependent variable. Specifically, you will learn

- how we transform the dependent variable;
- how we interpret the impact of the predictors after we have done the transformation;
- how to read a table of logistic regression results; and
- how we can extend logistic regression to deal with a dependent variable with more than two categories.

Why Logistic Regression?

If the dependent variable is dichotomous, sociologists usually employ logistic regression (LR) rather than the ordinary least squares (OLS) regression we have examined thus far. We no longer use the principle of least squares to define the equation, and the form of our dependent variable changes.

Suppose we want to predict criminal victimization. We do not try to predict whether someone has been victimized. Rather, we focus on the odds on being victimized. If, for example, two people out of three have been victimized, the odds on victimization are two to one. If only one has been victimized the odds are one to two (or, we would more likely say, two to one against). In defining our equations, we work with the natural log of the odds, so if you are part of the post-log generation, you may wish to consult Appendix B. An equation takes the form

$$\ln(\text{odds on a "1"}) = a + b_1 x_1 + b_2 x_2 + \ldots + b_n x_n,$$

Logits

The logged odds are typically referred to as "logits." We move the odds into this form because, unmodified, they have two undesirable features.

Suppose we have even odds on some outcome. These can be expressed as 1:1, or 1 / 1, or just 1.0. Suppose then we have odds of a thousand to one in favour of some other outcome, and odds of a thousand to one against a third. The favourable odds can be expressed as 1,000:1, or 1,000 / 1, or 1,000. The unfavourable odds can be expressed as 1:1,000, 1 / 1,000, or .001. Note how differently favourable and unfavourable odds are treated. Even odds are represented by the number one, favourable odds are represented by a number which differs by 999, and unfavourable odds are represented by a number which differs by .999. If we tried to minimize our errors in predicting odds, we would give FAR more weight to favourable than to unfavourable odds.

The second problem is that using odds can lead to predicted values that make no sense. The most unfavourable odds possible will be represented by a number not less than zero, but we can easily get predicted values below zero from a regression equation.

Logits solve both problems. The logged odds are positive for favourable odds and negative for unfavourable. They go as far from zero for favourable as for unfavourable odds, so we don't have to worry about underplaying the importance of unfavourable odds. As well, the logits can go to $\pm \infty$ (for probabilities of 1.00 and .00), so we can't get predicted values outside a range that makes sense.

The relationship between logits and probabilities is illustrated below. Figure 16.1 shows a few logit values not included in Table 16.1, so as to display the curve more clearly at the extremes.

Table 16.1: Probabilities and Corresponding Logits	
P	**Logit**
.01	−4.595
.05	−2.944
.10	−2.197
.20	−1.386
.30	−.847
.40	−.405
.50	.000
.60	.405
.70	.847
.80	1.386
.90	2.197
.95	2.944
.99	4.595

Note that the logit is 0 for p = .50, and that the logits are symmetric around .50—as p moves a given distance from .50, the logit changes by the same amount whether p is rising or falling. However, the logits for p < .50 are negative while those for p > .50 are positive.

Figure 16.1: Logits as a Function of ps

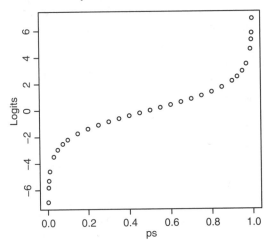

Interpreting the b's

In one sense, our standard interpretation for regression coefficients continues to hold. B_j gives us the change in the DV we predict for a one unit change in x_j, with other IVs held constant. In another sense, the interpretation is different because the change we expect is not in the value of y, but in the natural log of the odds on a "1."

For people not used to them, a change in logged odds may be awkward to interpret, so we often exponentiate our b's. Recall that to exponentiate is just the opposite of logging; in this case, we take e^{bj}. This gives us the factor by which we multiply the odds on a "1" for each one unit change in the value of x_j, other predictors held constant. So, if the exponential of $b_j = 1.5$, we multiply the odds on a "1" by 1.5 for a one unit change in x_j. Or, if $e^{bj} = .800$, we multiply the odds on a "1" by .8 for a one unit change in x_j. These multiplicative factors give us the ratio between the odds for someone with a coefficient of k and someone with a coefficient of k + 1, so they are often referred to as "odds ratios."

It is easiest to see why this is so by illustration. For simplicity, let us suppose that a predictor's initial value is 10, and that we want to see what will happen when it rises to 11. Let us suppose further that its exponentiated b is precisely 1.5, and that, for cases scoring 10 the odds on a "1" are 2:1, cancelling down to 2. As we move from cases scoring 10 to those scoring 11, we multiply the odds by exp(b), which is 1.5. Doing so gives us

$$2 \times 1.5 = 3$$

Now suppose we take the ratio of the odds for cases scoring 11 to those for cases scoring 10. We have

for the 11s, 3,
for the 10s, 2,
and the ratio is 1.5.

That is, the ratio of the odds for the higher category to those for the lower category equals the exponentiated b. Since many people find odds ratios more immediately meaningful than exponentiated b's, the expression "odds ratio" has become more and more common.

If we have a dichotomous predictor, all we need to define the effect of x_j is the exponential of b_j. Suppose we want to know how sex of respondent (coded 0 = M and 1 = F) influences the odds on theft of personal property, and suppose the exponential is .975. Then, as we move from the males to the females, we have to multiply the odds by .975. That is, being female ever so slightly lowers the odds.

Similarly, if we have a polytomous nominal variable, which we have converted to dummy variables, all we need to define the effect of a dummy is the exponential of its b. This gives us the factor by which we multiply the odds when we move from the reference category to the category singled out by the dummy.

If we have a predictor with a range of values—for example, education—then the exponential gives us the factor by which we multiply the odds for each one unit change along the scale. Let us suppose we have education coded in 11 categories, from elementary incomplete = 0 to graduate degree = 10. Let us suppose that the odds on theft of personal property are multiplied by 1.05 for each unit of change. Then the difference between one category and the next is not great: there will only be a 5% difference. But we have to multiply the odds once for each unit of change on the scale, so if we compare those with elementary incomplete to those with graduate degrees, we have to multiply 10 times: $1.05^{10} = 1.629$. That is, as we go from one end of the education scale to the other, the odds on victimization are multiplied by 1.629, or, in other words, the odds rise by 62.9%.

Converting Odds to Probabilities

For each case, we obtain a predicted value of ln(odds on a "1"). From these we can obtain predicted probabilities. Ordinarily we are happy to allow software to provide them, but we should know what is happening behind the scenes. We can get probabilities ourselves through a two-stage process. First we obtain the exponentials of our predicted values. The predicted values give us the predicted odds on a "1," in their logged form. When we exponentiate, we get the predicted odds on a "1," in unlogged form. Given the unlogged odds, we can go to probabilities from the formula

$$p = (odds) / (odds + 1)$$

So, for example, if we obtain predicted odds of .5 (that is, of 1:2, or two to one against), then

$$p = .5 / (.5 + 1) = .5 / 1.5 = .333$$

Given the ps for our cases, we can then, if we wish, produce graphs showing trends in the ps, such as the one in Figure 16.2.

The plot is based on results from DuMouchel's (1983) equation predicting self-reported use of narcotics for four cities. Including a dummy for Windsor, his equation was

$$-7.66 + .315*(Age) - .27*(Windsor)$$

For someone aged 16 and living in Windsor, this would come to

$$-7.66 + .315(16) - .27(1) = -7.66 + 5.04 - .27 = -2.89$$

Exponentiating (taking $e^{(-2.89)}$) gives odds of .0556. From

$$p = odds / (odds + 1) ,$$

we get .0556 / 1.0556 = .0527.

The other probabilities could be obtained the same way. These probabilities could easily have been graphed, but since many people are more used to percentages, the ps have been multiplied by 100, and the resulting percentages have been presented.

Figure 16.2: Percentage Reporting Narcotic Use, by Age and Site

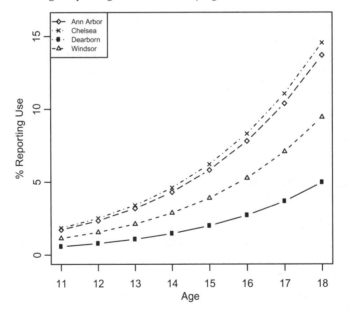

The lines are curved because taking products repeatedly produces an accelerating (exponential) trend. If we prefer a straight-line plot, we can plot the logs of the ps to get a straight trend line. Since many people are not used to thinking of the logs of probabilities, we will often be better to present curves.

They present a straightforward message: as age rises from 12 to 18, the probability of having used narcotics rises at an accelerating rate. The percentage reporting narcotic use varies from one city to another, but all show a curve of the same form.

For a second example, we shall take results from Hutchings' (2012) examination of the predictors of academic probation among first-year students at a Canadian university. As independent variables, Hutchings considered age and sex, living in residence, full- or part-time studies, faculty of enrolment, and incoming high school average. For those coming into university directly from Canadian high schools, her results implied the equation

12.70 − .19*(incoming average) + .20*(female) + .44*(residence) + .85*(Science) − 1.21*(Nursing)

The latter four variables were all dichotomies, all coded with the categories one unit apart, so the b's show the effect on the logits of moving from one category to another. Hutchings also obtained coefficients for other faculties, but for simplicity we shall consider only Nursing and Science, along with the reference Faculty of Arts and Social Science. Figure 16.3 shows the effect of incoming average, over the range from 75 to 90, and faculty.[1]

Figure 16.3: Percentage on Academic Probation, by Incoming Average

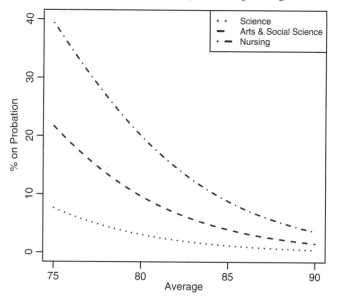

A Sample Logistic Regression Table

The variable to be predicted is Violent, coded 1 when someone reported having been the victim of a violent crime in the year before the General Social Survey of 2009, 0 otherwise. The predictors are as follows:

- Sex, coded 1= M, 2 = F;

- Married or widowed, coded 1 = married or widowed, 0 = neither;

- Age group, for which the lowest category is 15–19, and succeeding categories move up in five-year increments; and

- Evening_activities, coded 0–62.

Respondents were asked how often they did each of a series of things in the evening (going to work or school, going to pubs and bars, visiting friends and relatives, shopping, and so on). Their answers were summed and placed in Evening_acts.

Table 16.2 includes four columns, one to provide the names of the IVs, one for the b's, one for their exponentials, and one for p. The latter is the probability of getting a coefficient as extreme, or more extreme, through random fluctuations.

Table 16.2: Predictors of Violent Victimization

Predictor	b	exp(b)	p
Sex	−.077	.926	.714
Married or widowed	−.618	.539	.071
Age group	−.164	.849	.000
Evening activities	.017	1.017	.000
Constant	−1.896	.150	.000

There is little to be gained in graphing the results for the two dichotomous predictors. Neither is statistically significant, and the difference between two categories can more readily be grasped from exp(b) than a trend across multiple categories can be.

Figure 16.4 provides graphs for the two non-dichotomous predictors.

Figure 16.4: Effects of Age and Evening Activities

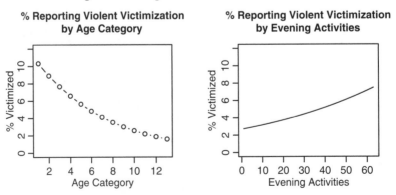

Graphing the results for these predictors shows their effect, over their full range of values, on a common scale. Over their full range, they each appear much more impressive than their exp(b)s by themselves would suggest. For age category, exp(b) is .849, but when we multiply the odds by this factor across several categories, then convert the odds to percentages, we see a striking drop. From the lowest category (15–19) to the highest (80+), the percentage reporting violence drops by a factor of 6.52. Similarly, the exp(b) for evening activities, at 1.017, by itself appears trivial, but multiplying the odds by this factor 62 times (moving from 0 to 62), and converting to percentages shows a strong effect: the percentage reporting violence rises by a factor of 2.72. Plainly, the impact of a predictor must not be judged simply by its coefficient.

An Extension of the Logistic Model: Multinomial Regression

An extension of logistic regression allows us to handle nominal polytomies. Such variables are common in the social sciences, including, for example, political party preference, types of crimes of which one might be the victim, and choices among programs of study. Because there are many such variables, the extension of logistic regression that deals with them has been widely applied. It has been referred to by varying names, including "multinomial logits," "multinomial logistic regression," and "multinomial regression," which we will use here.

Before it became readily available, response categories had to be examined two at a time. One way was to compare all others to a reference category. For example, people could be asked about being victims of a crime in the previous year, and could then be placed in four categories:

0 - no victimization;

1 - property crime;

2 - violent crime; and

3 - both.

Three logistic regressions could be used comparing those with no victimization to the other three groups, taken one at a time. (Any of the other categories could be used as a baseline point of comparison as well).

In social science journals today we much more often see multinomial regression. In this approach, we select a baseline (or reference) category, and then we obtain equations giving us the odds that a case is in the baseline or in another category, one equation for each other category. This sounds as though it could be done without multinomial regression software; however, the equations are estimated simultaneously rather than one at a time. One advantage is a gain in efficiency—in the long run, we will tend to get smaller standard errors this way.

We are also guaranteed that the probabilities that a case lies in each of the various categories will sum to one. Using separate equations to compare a baseline category to each of the others, we leave out cases in the remaining categories. When we use different cases in each equation, the probabilities do not sum to one across equations. In multinomial regression it is guaranteed that they will.

To illustrate the outcome of a multinomial regression analysis, Table 16.3 presents the results of an attempt to predict forms of criminal victimization. The reference group, coded 0, consists of those who reported no victimization in the previous year. The second group reported being the victims of a violent crime. The third group reported a household crime—e.g., breaking and entering or vandalism. The fourth reported both violent and household crimes, and the fifth reported other types of crime, most commonly theft of personal property.

Table 16.3: Predictors of Types of Criminal Victimization, Canada, 2009

| Predictor | Odds Ratio | | | |
	Violent Crime	Household Crime	Both	Other
Age group	.832**	.900	.777**	.866*
Evening activities—work and school	1.011*	1.008	1.021*	1.007*
Evening activities—other	1.031	1.013*	1.025***	1.020*

* $p < .05$; ** $p < .01$; *** $p < .001$

In tables of this kind, we may see the b's, which tell us how much the logit increases for one unit of change in a predictor. We may also see the exponentiated b's, which tell us how much we multiply the odds on a "1" for a one unit change in a predictor. Here, as is common today, the exponentiated b's are referred to as "odds ratios."

The predictors have been used in a previous example. The only difference is that evening activities have been split into those having to do with work and schooling and those involving other activities.

Let us walk through the table. In each equation the figure for age group lies below one, telling us that as we move from one five-year age category to the next we multiply the odds by a figure less than unity, and that therefore the predicted odds on a "1" decline with age. A similar kind of consistency applies to both forms of evening activity. The figure for each category of evening activity, in each equation, is greater than unity, so as evening activities increase so do the odds on each form of victimization.

Each of the predictors moves over a considerable range, so the cumulative effect of these multiplications is much more impressive than the reported odds ratios themselves. For example, there are 13 age categories, so as we move from the lowest (15–19) to the highest (80+), multiplying at each step, we multiply 12 times. In the case of violent victimization, for which the odds ratio is .832, the result is

$$(.832)^{12} = .110$$

The odds on victimization for those at the upper end of the age range are only 11% as great as those for 15–19 year olds.

For evening activities, not involving work or school, scores can range from 0 to 62, a score reached when someone, on average, carries out two activities every evening of the month. We can therefore, at the extreme, multiply the odds ratio 61 times. In the case of those who experienced both violent and household crime, the result is

$$(1.025)^{61} = 4.510$$

The odds on victimization are multiplied by just over 4.5 when we go from one extreme to the other. To assess the effect of a predictor, we must consider not just the size of its coefficient, but also the range over which it moves.

As we have seen, this point applies to logistic regression as well as its extension. A difference between the two lies in the statistical significance of the coefficients. There is no reason why the same predictor must be significant in all of the equations, or to the same level. Here only one predictor, age group, reaches .01 in two of the four equations, and it is not significant for household crime. Notice as well that only one predictor is significant for that form of crime, while all three work in the two equations on the right.

Because multinomial regression tables are so similar to those for logistic regression, no further material is needed to read them.

Summary

This final chapter has dealt with a special form of regression, which we use when the dependent variable is dichotomous. In logistic regression the DV cannot be used in its original form, but is transformed into a logit; that is, we work with the log of the odds on getting a "1." After the transformation, we can interpret the b's of logistic regression much like those of multiple regression: they give us the amount of change we expect in the DV for a unit of change in a predictor (other predictors controlled), but now the DV is a logit. More often now, we see reported not the b's, but their exponentials. These typically are referred to as odds ratios. They give us the factor by which we multiply the odds for a one unit change in a predictor (others controlled).

We have also seen a common extension of the logistic approach, in multinomial regression. We choose a baseline (or reference) category, and then we obtain equations to get the odds on being in each of the other categories, as opposed to the baseline. The equations are estimated simultaneously, so that the probabilities for a given case, across categories, always sum to one.

Review Questions on Logistic Regression

1. What is a natural logarithm? What is exponentiation?

2. Suppose we have taken the ln of a dependent variable, and we obtain the following equation:

$$ln(y) = 2.5 + .15(x1) + .07(x2)$$

How can we interpret the .15 and the .07?

3. What is a logit?

4. In logistic regression, what is the basic interpretation for b_j?

5. In logistic regression, we get a predicted logit for every case. Can we convert this to the probability of a one? How?

6. Suppose that in a survey of Canadians aged 15–64, ages are recorded in five-year categories, beginning with 15–19 and ending with 60–64. Suppose that in predicting violent criminal victimization, the b_j for age is .10. How much could the predicted value of the logit be affected by age?

7. If we wanted to show how much the unlogged odds on getting a "1" could be affected by age, what would we do?

8. Suppose that someone interested in property crime obtained data on victimization experiences and basic demographics from a sample. For this data set age was coded in five-year categories, and sex was coded 0 = M; 1 = F. Suppose the logistic regression results were as follows:

Predictor	b	p	exp(b)
Age	−.104	.004	.889
Sex	.054	.049	1.055
Constant	−6.325		

How much difference would there be in the unlogged odds for people who differ in age by 10 years? How much difference would there be between males and females?

9. What is the difference between logistic and multinomial regression?

10. What are two advantages of multinomial regression over doing a series of logistic regressions when our DV is polytomous?

Note

1. To create the graph, the values of "female" and "on campus" were set to their means. Thus the graphs show the effect of incoming average and faculty on students who were "typical" in this sense.

Appendix A: Going a Step Further

A1: Minimizing the Sum of Squared Deviations

To see that the sum of squared deviations reaches a minimum at $\bar{x}$, let us begin with the squared deviations from a general point k. Their sum is

$$\Sigma\,(x_i - k)^2\,,$$

or, by expansion of the squared term,

$$\Sigma\,(x_i^2 - 2kx_i + k^2) \qquad\qquad\qquad [1]$$

To show that this sum is at a minimum when $k = \bar{x}$, we consider the point $(\bar{x} + a)$ and show that we reach the minimum when $a = 0$.

Replacing k with $(\bar{x} + a)$ in [1], we have

$$\Sigma\,(x_i^2 - 2x_i(\bar{x} + a) + (\bar{x} + a)^2)\,,$$

which, after eliminating the central brackets becomes

$$\Sigma\,(x_i^2 - 2x_i\,\bar{x} - 2x_i\,a + (\bar{x} + a)^2)$$

After expanding the term on the right we have

$$\Sigma\,(x_i^2 - 2x_i\,\bar{x} - 2x_i\,a + \bar{x}^2 + 2a\,\bar{x} + a^2)$$

In finding a minimum value, we can ignore constants. Given the data, the x_i and $\bar{x}$ are constant, so we can drop x_i^2, $2x_i\,\bar{x}$, and $\bar{x}^2$, leaving

$$\Sigma\,(- 2x_i\,a + 2a\,\bar{x} + a^2)$$

Removing brackets leads to

$$- 2a\Sigma\,x_i + 2a\Sigma\,\bar{x} + \Sigma\,a^2$$

To sum the same value over N cases is to multiply by N, so we can replace Σa^2 with Na^2, giving us

$$-2a\Sigma\,x_i + 2aN\,\bar{x}\,\bar{x} + Na^2 \qquad\qquad\qquad [2]$$

By definition, $\bar{x} = \Sigma x_i\,/\,N$, so $N\,\bar{x} = \Sigma x_i$.

Replacing Σx_i in [2] with $N\,\bar{x}$ leaves

$$-2aN\bar{x} + 2aN\,\bar{x} + Na^2$$

Cancelling the first two terms, we are left to minimize

$$Na^2\,,$$

which takes its least value when $a = 0$, as required.

A2: The Mean and Standard Deviation of Z–Scores

We denote the mean of z as z-bar. It must be shown that

$$\bar{z} = 0$$

By definition,

$$\bar{z} = \Sigma\, z_i\, / \, N$$

Since z_i is defined as $(x_i - \bar{x})$ / $SD(x)$, we can substitute this expression for z_i.
    We get $\Sigma [(x_i - \bar{x}) / SD(x)] / N$, which can be written

$$\bar{z} = \frac{\Sigma\,(x_i - \bar{x})}{SD(x)(N)}$$

Above, it was shown that $\Sigma\,(x_i - \bar{x}) = 0$.
    So we have

$$\bar{z}\; \frac{0}{SD(x)(N)} = 0,$$

as required.
    It must also be shown that: $SD(z) = 1$.
    Squaring both sides yields

$$Var(z_i) = 1$$

From the definition of Variance, this can be written as

$$\Sigma\,(z_i - \bar{z})^2\, / \, N = 1$$

Using $\bar{z} = 0$, we have

$$\Sigma\,(z_i - 0)^2\, / \, N = 1 \text{ , or}$$

$$\Sigma\,(z_i)^2\, / \, N = 1$$

Using the definition of z_i, $(x_i - \bar{x})$ / $SD(x)$, we can substitute.

$$\Sigma\,([(x_i - \bar{x}) / SD(x)])^2\, / \, N = 1$$

Since $(SD(x))^2 = Var(x)$, this can be rewritten

$$\Sigma\,[(x_i - \bar{x})^2\, / \, Var(x)] \, / \, N = 1, \text{ or}$$

$$\frac{\Sigma\,(x_i - \bar{x})^2}{Var(x)(N)} = 1$$

    Recall that $Var(x) = \Sigma\,(x_i - \bar{x})^2\, / \, N$.
Since division by a fraction equals multiplication by its inverse, we can divide by Var(x) by
multiplying by

$$N / \Sigma (x_i - \bar{x})^2$$

Doing so gives us

$$\frac{\Sigma(x_i - \bar{x})^2(N)}{\Sigma(x_i - \bar{x})^2(N)} = 1, \text{ as required}$$

A3: The Standard Deviation of a Proportion

The standard deviation of a proportion is given by $\sqrt{p(1 - p)}$. The derivation is straightforward.

Let us code the two values of a dichotomy 0 and 1. In adding scores to get the mean, cases coded 0 make no contribution, so

$$\bar{x} = (\text{\# of 1s}) / N,$$

the proportion of 1s. If we denote this proportion as p, then the proportion of 0s is $(1 - p)$.

By definition,

$$\mathbf{Var(x)} = \Sigma (x_i - \bar{x})^2 / N$$

Substituting p for $\bar{x}$ yields

$$\Sigma (x_i - p)^2 / N \qquad\qquad\qquad \mathbf{[1]}$$

For cases where x = 0, $(x_i - p)^2 = (\bar{x} - p)^2 = p^2$.
Where x = 1, of course, it is $(1 - p)^2$.
Since there are only two values of x, we can rewrite [1] as

$$(\text{\# of 0s}) \ p^2 / N + (\text{\# of 1s})(1 - p)^2 / N$$

Using $(\text{\# of 0s}) / N = (1 - p)$, and $(\text{\# of 1s}) / N = p$, we have

$$(1 - p)(p)^2 + p(1 - p)^2$$

Expanding within the brackets on the right, we obtain

$$(1 - p)p^2 + p(1 - 2p + p^2)$$

Taking out the brackets on the right leads to

$$(1 - p)p^2 + p - 2p^2 + p^3$$

Taking out the brackets on the left gives

$$= p^2 - p^3 + p - 2p^2 + p^3$$

Cancelling gives us

$$p - p^2, \text{ which can be written } p(1 - p)$$

This is the Var(x). The SD is simply its root, $\sqrt{p(1 - p)}$. If we choose to denote $(1 - p)$ as q, we obtain an alterate version, $\sqrt{(pq)}$.

Appendix B: Some Additional Explanations

B1: Basic Notes on Logarithms

The basic idea of a logarithm is quite straightforward. Suppose we have an equation of the form

$$A = B^C$$

This just says that A is equal to some other number B, taken to a power C. For example, if

$$A = 100, B = 10, \text{ and } C = 2,$$

then

$$100 = 10^2$$

When we express A as equal to B, raised to the power C, we refer to B as the base, and to C as the logarithm. Here 10 is the base and 2 the logarithm.

Ten is a widely used base. We refer to the associated logarithms as common logarithms, or common logs for short. Consider some further examples:

$$10,000 = 10^4$$

$$1,000 = 10^3$$

$$100 = 10^2$$

$$10 = 10^1$$

$$1 = 10^0$$

$$.1 = 10^{-1}$$

$$.01 = 10^{-2}$$

As the exponent, the common logarithm, goes down by one, the number is divided by 10. Conversely, as it goes up by one, the number is multiplied by 10.

When we take the common log of a number, we just replace the number itself with the exponent to which 10 must be raised to produce that number. If we take the log of 1,000, we replace it with 3, or if we take the log of .01, we replace it with −2.

This might be expressed by saying

$$\log_{10}(1,000) = 3, \log_{10}(.01) = -2$$

The subscript after "log" just indicates the base for the logarithm. Outside of statistics and computer science, if you don't see a subscript, or a number in brackets, the base is usually 10.

In computing science and information theory, the base 2 is common. Here are the results of raising this base to various powers.

$$2^1 = 2$$

$$2^2 = 4$$

$$2^3 = 8$$

$$2^{-1} = \frac{1}{2} = .5$$

$$2^{-2} = \frac{1}{4} = .25$$

$$2^{-3} = \frac{1}{8} = .125$$

The logs are just the exponents to which the number 2 must be raised to get the original number.

Original number	Log_2
2	1
4	2
8	3
.5	−1
.25	−2
.125	−3

The most widely used base in statistics is the artificial number e. Since it is not an integer, being approximately equal to 2.7178, logs to this base cannot be calculated as simply as logs to the base 10 or 2. We can illustrate, though. We will, as above, place the base of the logs in a subscript. Note, though, that you will usually see "ln" used to refer to logs to the base e. These logs are often referred to as "natural logs," and the two letters in "ln" are just the initial letters of "log" and "natural." (In Latin, in which natural logs were first written about, "log" came before "natural.")

$$\log_e(100) = 4.605$$

$$\log_e(105) = 4.654$$

$$\log_e(110) = 4.700$$

Notice that, as the number whose log is taken (100) rises by 5% (to 105), the natural log rises by about .05. Again, as we go from 100 to 110, the natural log rises by about .10. These are illustrations of the general principle that, up to a percentage change of about 20, the natural log rises by about .01 for each percentage point of change. The principle is illustrated graphically below.

In Figure B.1, x rises from 100 to 120, a change of 20%. Ln(x) rises from a bit over 4.6 to a bit under 4.8, somewhat less than .20, but reasonably close to a rise of .01 for each percentage point of change.

Figure B.1: ln(x) versus x

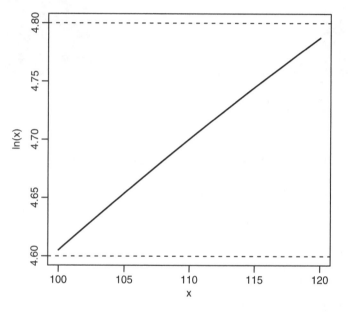

Exponentials

From logs we can move back to numbers on the arithmetic scale. Suppose we have a common log (one to the base 10) of 3. Ten (the base) to the power of 3 (the log), yields 1,000. If the log to the base of 2 is 3, we take 2 to the power of 3 and obtain 8. To move from a log to a number on the arithmetic scale this way is called "exponentiating," or "taking the exponential."

Logging to Change the Shape of a Distribution

If we take the logs of all values in a distribution, some will be affected more than others. As long as the base is greater than one (as it always is in the social sciences), extreme high scores will be reduced more than moderate or low scores. Thus outlying high scores will be moved closer to the bulk of the distribution. This effect is illustrated in Figure B.2 below, in which household incomes become much more symmetrical after taking the log.

Figure B.2: Original and Logged Distributions of Household Income

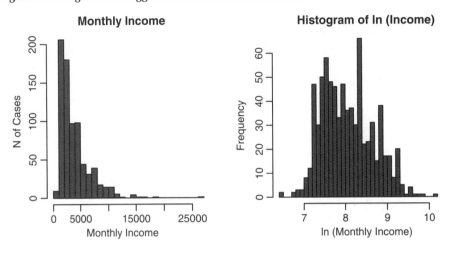

B2: Obtaining Expected Values for Chi-Square

For chi-square, we need the expected values for variables that are independent in the population from which we are sampling. We just need to see how to get them.

In probability theory, variables are independent if

$$p(i,j) = p(i)p(j) \, ,$$

where p(i,j) = the probability of i and j both occurring,

p(i) = the probability of i occurring, and

p(j) = the probability of j occurring.

For a crosstabulation, this means that

the probability a case is in the ith row and the jth column

equals the product of

the probability of its being in the ith row, times

the probability of its being in the jth column.

Suppose we have two rows and two columns, and that

the probability of being in row 1, p(i = 1) = .6,
the probability of being in row 2, p(i = 2) = .4,

the probability of being in column 1, p(j = 1) = .6, and
the probability of being in column 2, p(j = 2) = .4

Then

$$p(1,1) = (.6)(.6) = .36 ,$$

$$p(1,2) = (.6)(.4) = .24 ,$$

$$p(2,1) = (.4)(.6) = .24 , \text{and}$$

$$p(2,2) = (.4)(.4) = .16$$

We can place the p(i,j)s in a table:

.36	.24
.24	.16

To move from the expected probabilities to the expected number of cases in the cells we just multiply the p(i,j)s by the table total, denoted by N in the formula

Np(i,j),

from which we get the expected values. For a table with N = 40, we have

40(.36)	40(.24)
40(.24)	40(.16)

Multiplying through, the table of expected cell counts becomes

14.4	9.6
9.6	6.4

In practice, we are not usually given the p(i)s or the p(j)s, and we have to estimate them from the data. For the ith row, we take

R(i) / N ,

where R(i) is the row total for row i, and

N is the table total.
Similarly, for the jth column, we take

C(j) / N ,

where C(j) is the column total for column j.
Using this notation, to get Np(i,j), we take

$$N \times \frac{R(i)}{N} \times \frac{C(j)}{N} = \frac{R(i)C(j)}{N}$$

Examples of the use of this formula have appeared above.

B3: Another Form of Bayesian Hypothesis Testing

Another form of Bayesian test compares Bayes factors for competing models. (Recall that the Bayes factor is the quantity by which we multiply the prior to get the posterior.) Suppose we have a regression equation including x_k, and another without it. To test the hypothesis that x_k affects y, we can compare the Bayes factors for the two equations. If the ratio of Bayes factors suggests that the effect of x_k is not zero, we include it.

We are, in effect, comparing the likelihood that the effect of x_k is b_k (as it is when x_k is included) with the likelihood that b_k is zero (as it is when x_k is excluded). Instead of asking how often we would get a result of the kind we have (or more extreme) if a null hypothesis were true, the Bayesian approach compares two alternatives directly.

For sample sizes typical in sociology, comparison of Bayes factors is more stringent: a coefficient significant at .05 under the standard system may well be less plausible than a coefficient of 0 under the Bayesian. This greater stringency may reduce the frequency of non-reproducible results. (Or prevent us from picking up something of interest.)

Comparing Bayes factors does not necessarily require us to have very clear priors (although if we can justify them, so much the better). Whether vague or precise, we can combine them with data to develop our posteriors.

Glossary of Statistical Terms

analysis of covariance: a form of multiple regression, in which some predictors are categorical and others are not.

ANCOVA: an abbreviation for "analysis of covariance."

analysis of variance: a method to see how much one or more categorical predictors affect an outcome. It is a special case of regression, widely used in experiments.

ANOVA: an abbreviation for "analysis of variance."

ASH: an abbreviation for "average shifted histogram." To create one we use narrow bins, and shift the origin for the graph gradually. We average the number of cases at each location, and plot the averages.

association plot: a plot showing which cells of a table are heavy and which light, and thus showing us how two (or more) variables are associated.

b: stands for a regression coefficient, which will be the number of units of change in y we expect for a unit of change in a predictor.

back-to-back histogram: a graph in which histograms for two groups are plotted on the vertical axis, one on the left and one on the right, so the bars for the two groups are at the same levels.

bar chart: a graph in which the length of bars represents the frequency (or proportion) of cases in a category. The bars are separated to emphasize the distinctness of the categories.

base: in $A = B^C$, the base is B, the number which is raised to a specific power, here denoted C, so the right hand side will equal A. Commonly used bases are 2, 10, and e.

Bayes factor: the factor by which we multiply a prior probability to get a posterior probability.

Bayes' theorem: a theorem which shows how to combine prior probabilities with data to obtain updated (posterior) probabilities.

beta: in regression, "betas" are the coefficients we obtain when both y and its predictors have been standardized.

beta distribution: the beta distribution is one often used to represent the prior and posterior distribution of a proportion.

bias: a tendency for a measurement to be higher or lower than the true figure.

bimodal: having two pronounced peaks.

bit: a binary digit, which can be used to represent a unit of information.

bivariate regression: a way of predicting scores on one variable from those on another, in which the link between them is represented by a trend line, typically a straight line.

boxplot: a graph in which horizontal lines represent quartiles, and vertical lines descend from the upper quartile to the lower, forming a box which contains the middle half of the distribution. These plots also show the median, and how far scores lie above and below the outer quartiles.

broadened median: the mean of the median and a small number of points above and below it.

cell blocking: in log-linear models, treating cells of a table as though they do not exist.

Central Limit Theorem: a theorem showing that the sampling distribution of the mean becomes normal as sample size increases. It holds for any variable with definable variance.

centroid: the point at which two or more variables take on their mean values.

chi-square distribution: a distribution commonly used in statistical inference. The simplest chi-square distribution is created by squaring the values of observations in a normal distribution. If we square the values of random observations from more than one normal distribution, then add the scores, we get other chi-square distributions.

chi-square test: a test of statistical significance based on the chi-square distribution.

coefficient of determination: another name for r^2 or R^2. These measures tell us the proportion of variance in y accounted for by its predictors, or in other terms, to what extent y is determined by them.

coefficient of variation: the standard deviation of a variable divided by its mean.

common logarithm: a logarithm using the base 10.

concordant: a concordant pair of cases is one giving evidence of a positive association.

conditional independence: exists when two variables are unrelated, once we have taken another (or others) out of the picture.

conditional table: a crosstabulation including only cases which meet some condition, e.g. that they be male or female.

confidence band: in regression, a band within which a given percentage of trend lines can be expected to lie if we calculate them for a great many samples.

confidence interval: a range within which a given percentage of sample results can be expected to lie, over the long run.

continuous: a continuous variable is one which takes on all values within its possible range. Strictly continuous variables are theoretical rather than observed, although we sometimes speak of variables as continuous if they could, on principle, be observed at any value within their range.

covariance: a measure of how quantities vary together, or c̲o̲vary. A formula, which works when we know the means of both variables, is $\Sigma(x_i - \bar{x})(y_i - \bar{y})$ / N.

covariate: a predictor added to a regression equation (or an ANOVA) to control for its influence on the dependent variable.

credible interval: in Bayesian inference, a range within which the true figure is likely to lie, with a specified probability.

critical region: an area in the tail of a sampling distribution. An observed result which falls into the region will be taken as statistically significant.

deciles: points dividing an ordered distribution into tenths.

degrees of freedom: the number of values that are free to vary. It is often abbreviated as "df." The number of df is given by differing formulae for different tests, e.g. for the chi-square test of independence it is given by $(r - 1)(c - 1)$.

dependent variable: a variable to be predicted by others, or which is seen as caused by others.

DF: an abbreviation for "degrees of freedom."

dichotomy: a variable that takes on only two values.

direct effect: in path analysis, an effect on a dependent variable that is exerted without an intermediate variable.

discordant: a discordant pair of cases is one giving evidence of a negative association.

discrete: a discrete variable has distinct categories, which do not shade into one another.

discrete-continuous: a discrete-continuous variable has distinct categories, but these fall along an ordered range and may be quite close to one another.

dispersion: another term for variability or variation.

distortion: exists when the relationship between two variables is reversed when we control for a third.

dummy variable: a dichotomous variable we create, often from a categorical variable whose categories we wish to examine separately.

d: a measure of association for ordinal variables, proposed by Somers, ranging from −1.00 to 1.00. It is based on concordant and discordant pairs, and pairs tied on the dependent variable but not the independent.

e: an artificial number approximately equal to 2.7178.

entropy: lack of organization.

error bars: bars placed around an estimate in a graph to show the standard error of the estimate.

Error Sum of Squares: the sum of squared differences between the observed and predicted values of a dependent variable.

Error Variance: the part of the variance in y that we cannot account for with our set of predictors. Its numerator is the Error Sum of Squares.

exponential(b): in logistic regression, the factor by which we multiply the unlogged odds on a "1" for a one unit change in a predictor. Often referred to as the "odds ratio."

exponential curve: an accelerating (or decelerating) curve, created when values of one variable increase (or decrease) by a constant percentage.

exponentiate: to move back from a logarithm to a number on the arithmetic scale. If $A = B^C$, and we have C, the logarithm, to exponentiate C is to raise B to the power of C. This process is the opposite of taking the logarithm of A to the base B.

frequentist probabilities: probabilities based on long-run frequencies of events.

gamma: a measure of association for ordinal variables, ranging from −1.00 to 1.00, and based on concordant and discordant pairs.

G-squared: another name for "L-squared" in log-linear modelling. Under the null, this quantity is distributed as chi-square, and so can be used in tests of fit.

graph of averages: a graph in which, for a given value of x, we plot the average value of y.

histogram: a graph in which the length of bars represents the number (or proportion) of cases in a category. Histograms are used with ordered variables for which we do not wish to emphasize the distinctness of categories by putting spaces between the bars.

IDR: an abbreviation for "interdecile range."

independence: inability to predict one variable from another.

independent variable: one used to predict another, or viewed as a cause of the other.

index of diversity: a measure for nominal data, which gives us the probability that two cases chosen at random will be in different categories.

Index of Qualitative Variation: a measure for nominal data, which gives us the unmatched pairs as a proportion of the greatest possible number of unmatched pairs.

indirect effect: in path analysis, an effect on a dependent variable that is exerted through an intermediate variable.

inefficient: in sampling, refers to a method that requires more cases to obtain as much precision as could be had through another (more efficient) method.

intercept: the point at which a line passes through the y-axis of a graph.

interdecile range: the distance between the upper and lower deciles (D_9 and D_1) of an ordered distribution.

interquartile range: the distance between the upper and lower quartiles (Q_3 and Q_1) of an ordered distribution.

interval: one of Stevens' levels of measurement, in which categories are ordered and we know the precise distances between them.

IQR: an abbreviation for "interquartile range."

Ku: one of Pearson's measures of kurtosis, based on $(x_i - \bar{x})^4$.

lambda: a measure of association for nominal variables, which gives us the proportion by which we can reduce our errors in guessing the category a case is in if we know where the case stands on another variable.

leptokurtic: having a high central peak and long tails.

ln: stands for a natural logarithm, one to the base e.

logarithm: in $A = B^C$, C is the logarithm, the power to which B must be raised so the right-hand side equals A.

logit: in logistic regression, the logged odds on getting a "1."

L-squared: a quantity used in testing log-linear models, which is distributed as chi-square when the null hypothesis is true. Since it is distributed as chi-square, it can be used in tests of fit. Also referred to as "G-squared" and "the likelihood ratio chi-square."

mean: the standard arithmetic average, obtained by summing scores and dividing by the number of cases.

median: the point which divides an ordered distribution into lower and upper halves.

mesokurtic: having a moderate central peak and tails.

metric coefficients: regression coefficients obtained without standardizing the variables.

mode: as a measure of central tendency, the most common category; in examining a distribution, a pronounced peak in the graph.

mosaic plot: a plot in which rectangles represent cells of a table, and are proportional in area to the number of cases in the cells.

moving average: consists of averaged scores for adjacent points in time, the points used shifting as we move across a graph.

multi-modal: having more than two pronounced peaks.

multiple regression: a way of predicting scores on one variable from those on a set of others, in which the link between the dependent variable and each of the others is represented by a trend line, usually a straight line.

mutual independence: in log-linear modelling, a situation in which no variable can be predicted from another.

natural logarithm: a logarithm taken to the base e.

negative association: an association in which as one variable rises the other falls.

nominal: one of Stevens' levels of measurement, in which categories have no meaningful order, so that numbers assigned to them are arbitrary.

normal distribution: a continuous, unimodal, and symmetric distribution widely used in inferential statistics. When graphed, it is often referred to as a "bell curve."

null hypothesis: typically the hypothesis that there is no difference between groups or no association between variables. It is contrasted with a research hypothesis.

odds ratio: simply the ratio of two odds. In a 2 × 2 table, with cells labelled

    a   b
    c   d

the odds on being in row one, for column one, are just a / c. The odds for column two are b / d. As can be shown with a little algebra, the ratio of these is ad / bc.

one-tailed test: a test of a directional hypothesis, e.g. that one group will score higher than another or that an association will be positive.

one-way ANOVA: a form of analysis of variance in which there is only one independent variable (which may have several categories).

ordinal: one of Stevens' levels of measurement, in which the categories are ordered, but we do not know the precise distances between them.

p: an abbreviation for "probability."

p-value: the probability of obtaining a result of the size we have, or a more extreme one, by chance.

partial Q: Q calculated from conditional tables that remove the effect of a third variable.

Pearson's r: the most common measure of association for interval or ratio variables. Often simply called "r."

percentiles: points dividing an ordered distribution into 100 categories with an equal number of cases in each.

personal probabilities: probabilities estimated by an individual, based on whatever evidence and arguments are seen as helpful.

phi: a measure of association for two dichotomous variables. Numerically equal to Pearson's r for the same data.

platykurtic: having a flat central peak and short tails.

population pyramid: a type of back-to-back histogram, in which males are on one side and females on the other, and the horizontal bars represent age categories.

positive association: an association in which as one variable rises so does the other.

posterior probabilities: probabilities obtained by combining prior probabilities with data, to get a revised view of how things probably are.

power: the probability that a statistical test will detect an effect of a given size, if one is present in the population from which a sample is to be drawn.

PRE: stands for "Proportional Reduction in Error."

prior probabilities: probabilities estimated by an individual before fresh data are obtained.

Q: a measure of association for 2 × 2 tables, numerically equal to gamma for tables of this size.

quadratic curve: the curve resulting when a variable is a function of another whose values have been squared. A pure quadratic will have one point of inflection.

qualitative measures: these include variables for which arithmetic is not logically possible. Stevens' nominal and ordinal variables count as qualitative.

quantitative measures: these include variables with which arithmetic can be done. This requires that intervals between categories be well defined. Stevens' interval and ratio measures qualify as quantitative.

quartiles: points which divide an ordered distribution into quarters.

quasi-independence: in log-linear modelling, this refers to independence between variables after we have blocked cells.

r: a measure of association for interval or ratio variables, ranging from −1.00 to 1.00, often referred to as "Pearson's r" after its originator.

regression coefficient: commonly referred to as b, gives us the number of units of change we expect in y for a unit of change in an independent variable.

Regression Sum of Squares: the sum of squares that is accounted for by the predictors in regression. The RSS equals the difference between the total sum of squares and the error sum of squares.

research hypothesis: a hypothesis whose plausibility we would like to check. Contrasted with the null hypothesis.

residuals: differences between the values predicted under a model and those observed.

resistant: not influenced (greatly) by outlying cases.

robust: not influenced greatly by modest failure of underlying assumptions.

sampling distribution: the distribution of results we would have if we calculated a statistic for each of an infinite number of samples and stored the results.

saturated model: in log-linear work, a model in which every possible effect has been included.

scatterplot: a graph in which cases are placed at points corresponding to their scores on two variables, one plotted on the x-axis and the other on the y-axis.

SD: an abbreviation for standard deviation.

SE: an abbreviation for "standard error," the standard deviation of a sampling distribution.

SEE: an abbreviation for standard error of estimate, the standard deviation of the residuals from a regression equation (or ANOVA).

$s_{y.x}$: notation for the standard error of estimate, indicating that y is predicted by x.

$s_{y.x1-xn}$: notation for the standard error of estimate, indicating that y is predicted by x1 through xn.

significance test: a test to see whether our observed results, or more extreme results, would come up often by chance. If they would arise infrequently, we say the observed results are statistically significant.

Sk: one of Pearson's measures of skewness, based on $(x_i - \bar{x})^3$.

slope: a term sometimes used to refer to the regression coefficient, since, if we plot the trend line on a scatterplot, its slope is given by the coefficient.

specification: exists when the association between two variables is different for subsamples with different values of a third variable.

spline: in regression, a second variable created from an independent variable so that a change in the slope can be dealt with. Typically, the spline is set up so its b represents the difference between the slope on the left and the slope on the right.

spurious association: one in which the observed correlation between two variables exists because each is affected by a common cause.

stacked bar chart: a bar chart in which the bars are divided into bands showing the size of sub-categories.

standard deviation: a measure of variation for interval and ratio variables.

standard error: the standard deviation of a sampling distribution.

standard error of estimate: the standard deviation of the residuals from a regression model.

standardized residuals: differences between predicted values under a model and the observed values, if divided by their standard errors, are called standardized residuals. Often used to find unusually heavy or light cells in crosstabulations.

standardized variable: one in which the mean value is subtracted from each observed value, and the remainder is divided by the standard deviation of the variable. Scores for individuals on standardized variables are referred to as z-scores.

suppression: when the link between two variables appears much weaker before a third variable is controlled for, we speak of suppression.

t distribution: a continuous, unimodal and symmetric distribution widely used in inferential statistics. It has a lower central peak and heavier tails than the normal distribution, but as sample size increases approaches the normal.

test factor: in checking to see whether the link between two variables depends on the value of a third, the third may be referred to as a test factor.

three-factor interaction: three-factor interaction exists when the value of one variable is affected by the combined values of the other two.

Total Sum of Squares: the sum of squared differences between individual observations and their mean.

trimmed mean: a mean calculated with a percentage of cases trimmed away from both the upper and the lower ends of the distribution.

truncation: restricting the range of a variable, by recoding values above (or below) a specific value to equal that value.

two-tailed test: a test of a non-directional hypothesis, that one group will score either higher or lower than another, or that two variables will be either positively or negatively associated.

two-way ANOVA: ANOVA with two independent variables.

two-way effect, interaction: in log-linear work, an association between two variables.

Type I error: abandoning the null hypothesis when it is true.

Type II error: failing to abandon a null hypothesis when it is false.

unanalyzed effect: a link between a dependent variable and a predictor that is not examined causally. Variables along the path between them have unclear causal status, or the analyst may have no need to assess the link in causal terms.

uncertainty coefficient: a measure of association for nominal variables, based on measures of entropy.

unimodal: displaying a single pronounced peak.

V: a measure of association for nominal variables, due to Cramèr, based on chi-square and set up so that it will lie between 0 and 1 for two-way tables of any size.

x-bar, or $\bar{x}$: a symbol for the mean of the variable x.

y-hat, or $\hat{y}$: the predicted value of the dependent variable in regression.

z-score: individual values of a standardized variable are often referred to as z-scores.

Credits

Table 12.3: Admission by Sex. **Figure 12.2:** Admission by Sex by Program. Based on data from P.J. Bickle, E.A. Hammel, and J.W. O'Connell. (1975). "Sex Bias in Graduate Admissions: Data from Berkeley." *Science*, 187(4175):398–404.

Table 12.7: Mortality by Smoking Status. **Figure 12.3:** Mortality by Smoking and Age. **Table 12.8:** Mortality by Smoking Status and Age at Initial Interview. **Figure 12.4:** Effects of Age and Smoking on Mortality. **Table 12.9:** Probability of Death by Smoking Status and Age. Adapted from David R. Appleton, Joyce M. French, and Mark P.J. Vanderpump (1996). "Ignoring a Covariate: An Example of Simpson's Paradox." *The American Statistician*, 50(4):340–41.

Table 12.10: Contraceptive Practice by Socio-economic Category. **Figure 12.5:** Contraception by Socio-Economic Category for Those Born Outside North America. **Table 12.11:** Contraceptive Practice by Socio-economic Category, for Those Born Outside North America. Adapted from Robert Arnold, Cyril Greenland, and Marylen Wharf (1974). *Family Planning in Hamilton*. Planned Parenthood Society, Hamilton, ON.

Figure 13.1: Heights of Offspring by Mid-heights of Parents, in Inches. **Figure 13.2:** Median Height of Child by Mid-height of Parents. Adapted from data in "UsingR" and "psych" packages for R (R Core Team [2013] *R: a Language and Environment for Statistical Computing*. R Foundation for Statistical Computing, Vienna. url = www.r-project.org).

Figure 13.10: BMI by Age. Rebecca Draisey (2013). Predictors of BMI in Cambodia, and Investigation of Central Obesity in Rural Cambodians in Association With Diabetes and Hypertension. Unpublished MASDA major paper, University of Windsor.

Figure 13.13: Canadian Population in Millions, 1861–2011. **Figure 13.14:** ln(Canadian Population) 1861–2011. Based on data from the Census of Canada.

Table 14.3: Determinants of Support for Non-marital Cohabitation, for Unmarried Respondents. **Figure 14.7:** Job Satisfaction by Blishen Score and Age. Based on data from the All Alberta Survey 2008.

Figure 14.6: Depression Score by Years of Education. Adapted from Catherine E. Ross and John Mirowsky (2006). "Sex differences in the effect of education on depression: resource multiplication or resource substitution?" *Social Science and Medicine* 63:1400–13.

Figure 14.12: Effect of Statement About Greater Male Ability on Self-Rating of Task Performance. Adapted from Shelly J. Correll (2004). "Constraints into Preferences: Gender, Status, and Emerging Career Aspirations." *American Sociological Review*, 69:93–123.

Figure 14.13: Effects of F's Style of Dress and M's Availability on N of Status Products Recalled. Based on data from K. Janssens, B. Pandelaere, and M. Van den Bergh (2010). Can buy me love: Mate attraction goals lead to perceptual readiness for status products. *Journal of Experimental Social Psychology*, 47:254–58.

Figure 15.1: Blau and Duncan's Model of Occupational Attainment. Adapted from Peter M. Blau and Otis Dudley Duncan. (1967) *The American Occupational Structure*. New York: Wiley.

Figure 16.2: Percentage Reporting Narcotic Use, by Age and Site. Adapted from William H. Dumouchel (1976). On the Analogy Between Linear and Log-linear Regression. Technical Report 67, Dept. of Statistics, University of Michigan.

Figure 16.3: Percentage on Academic Probation by Incoming Average. Adapted from Krystin Hutchings (2012). "Academic Probation: First Year Students in Academic Studies." Unpublished MASDA major paper, University of Windsor.

References

Appleton, David R., Joyce M. French, and Mark P.J. Vanderpump. 1996. Ignoring a Covariate: An Example of Simpson's Paradox. *The American Statistician*, 50(4):340–1.

Arnold, Robert. 1972. *The Hamilton Study of Poverty*. Social Planning and Research Council of Hamilton and District, Hamilton ON.

Arnold, Robert, Cyril Greenland, and Marylen Wharf. 1974. *Family Planning in Hamilton*. Planned Parenthood Society, Hamilton ON.

Arnold, Robert, Michael Wheeler, and Frances Pendrith. 1980. *Separation and After: a Research Report*. Ontario Ministry of Community and Social Services, Toronto ON.

Anscome, F.J. 1973. Graphs in Statistical Analysis. *The American Statistician*, 27(1):17–21.

Bickle, P.J., E.A. Hammel, and J.W. O'Connell. 1975. Sex Bias in Graduate Admissions: Data from Berkeley. *Science*, 187(4175):398–404.

Blau, Peter M., and Otis Dudley Duncan. 1967. *The American Occupational Structure*. Wiley, New York.

Cleveland, William S., and Robert McGill. 1985. Graphical Perception and Graphical Methods for Analyzing Scientific Data. *Science* 229(4716): 828–33.

Cleveland, William S., and Robert McGill. 1987. Graphical Perception: The Visual Decoding of Quantitative Information on Graphical Displays of Data. *Journal of the Royal Statistical Society. Series A (General)*, 192–229.

Correll, Shelley J. 2004. Constraints into Preferences: Gender, Status, and Emerging Career Aspirations. *American Sociological Review*, 69:93–123.

Draisey, Rebecca 2013. Predictors of BMI in Cambodia, and Investigation of Central Obesity in Rural Cambodians in Association with Diabetes and Hypertension. Unpublished MASDA major paper, University of Windsor.

Dumouchel, William H. 1976. On the Analogy between Linear and Log-linear Regression. Technical Report 67, Dept. of Statistics, U. of Michigan.

Hamilton, Richard, and Maurice Pinard. 1976. The Bases of Parti Québécois Support in Recent Quebec Elections. *Canadian Journal of Political Science*, 9(1):3–26.

Hutchings, Krystina. 2012. Academic Probation: First-Year Students in Academic Studies. Unpublished MASDA major paper, University of Windsor.

IBM Corp 2012. *IBM SPSS Statistics for Windows*, Version 21.0. IBM Corp., Armonk, New York.

Janssens, K., B. Pandelaere, and M. Van den Bergh 2010. Can Buy Me Love: Mate Attraction Goals Lead to Perceptual Readiness for Status Products. *Journal of Experimental Social Psychology*, 47:254–8.

Nakhaie, R., and R. Arnold. 2010. A Four-Year (1996–2000) Analysis of Social Capital and Health Status Of Canadians: The Difference That Love Makes. *Social Science and Medicine*, 71(50):1037–44.

Peters, R. DeV., K. Petrunka, and R. Arnold. 2003. The Better Beginnings, Better Futures Project: A Universal, Comprehensive, Community-Based Prevention Approach for Primary School Children and Their Families. *Journal of Clinical Child and Adolescent Psychology*, 32(2):215–27.

Peters, Ray Dev., Alison J. Bradshaw, Kelly Petrunka, Geoffrey Nelson, Yves Herry, Wendy M. Craig, Robert Arnold, Kevin C.H.Parker, Shahriar R. Khan, Jeffrey S. Hoch, S. Mark Pancer, Colleen Loomis, Jean–Marc Bélanger, Susan Evers, Claire Maltais, Katherine Thompson, and Melissa D. Rossiter. 2010. The Better Beginnings, Better Future Project: Findings from Grade 3 to Grade 9, *Monographs of the Society for Research in Child Development*, 75(3):1–176.

R Core Team. 2013. R: A Language and Environment for Statistical Computing. R Foundation for Statistical Computing, Vienna. www.r-project.org

Raftery, Adrian. 2002. Statistics in Sociology 1950–2000, pp. 156–70 in Adrian Raftery,

Martin A. Tanner, and Martin T. Wells (eds), *Statistics in the 21st Century*, Chapman and Hall, New York.

Robbins, Naomi B. 2005. *Creating More Effective Graphs*. Wiley, New York.

Ross, Catherine E., and John Mirowsky. 2006. Sex Differences in the Effect of Education on Depression: Resource Multiplication or Resource Substitution? *Social Science and Medicine* 63:1400–13.

Shannon, C.S. 1948. A Mathematical Theory of Communication. *Bell System Technical Journal*, 27:379–423, 623–56.

StataCorp. 2013. *Stata/SE 12.1*. StataCorp LP, College Station, Texas.

Stevens, S.S. 1946. On the Theory of Scales of Measurement, *Science*, 103:677–80.

Stevens, S.S. (Ed.). 1951. *Handbook of Experimental Psychology*. Wiley, New York.

Tukey, John W. 1961. Cited in Velleman and Wilkinson 1993.

Tufte, Edward R. 1983. *The Visual Display of Quantitative Information*. Graphics Press, Cheshire CT.

Velleman, Paul, and Leland Wilkinson 1993. Nominal, Ordinal, Interval, and Ratio Typologies Are Misleading. *The American Statistician*, 47(1):65–72.

Vysochanskij, D.F., and Y.I. Petunin. 1980. Justification of the 3 Σ Rule for Unimodal Distributions. *Theory of Probability and Mathematical Statistics* 21:25–36.

Index

Page numbers in *italics* indicate figures.